CASE STUDIES IN ALLERGIC DISORDERS

CASE STUDIES IN ALLERGIC DISORDERS

Hans Oettgen • Raif Geha
Harvard Medical School

NEW YORK AND LONDON

Vice President: Denise Schanck
Senior Editor: Janet Foltin
Development Editor: Eleanor Lawrence
Senior Editorial Assistant: Allie Bochicchio
Production Editor: Ioana Moldovan
Typesetter and Senior Production Editor: Georgina Lucas
Copy Editor: Bruce Goatly
Proofreader: Sally Huish
Illustrations and Cover Design: Matthew McClements, Blink Studio, Ltd.
Indexer: Medical Indexing Ltd.

ISBN 978-0-8153-4436-0

Library of Congress Cataloging-in-Publication Data
Oettgen, Hans.
Case studies in allergic disorders / Hans Oettgen and Raif Geha.
p. ; cm.
ISBN 978-0-8153-4436-0 (alk. paper)
I. Geha, Raif S. II. Title.
[DNLM: 1. Hypersensitivity--Case Reports. 2. Immune System Diseases--Case Reports. WD 300]

616.9709--dc23

2012038405

Published by Garland Science, Taylor & Francis Group, LLC, an informa business
711 Third Avenue, 8th Floor, New York, NY 10017, USA and
3 Park Square, Milton Park, Abingdon, OX14 4RN, UK.

Printed in the United States of America

15 14 13 12 11 10 9 8 7 6 5 4 3 2 1

Visit our Website at http://garlandscience.com

Preface

In 1949 an exasperated U.S. Air Force Captain named Ed Murphy was trying to sort out wiring errors in an electrical circuit, when he uttered the now infamous words "*anything that can go wrong, will go wrong.*" Like his electrical circuit, the immune system is a complex network of interacting bits. Every day, clinicians experience the reality of Murphy's law: just about anything that can go wrong with the immune network actually will. In this case, these errors are myriad cellular and molecular dysfunctions that manifest as immunological disease.

This book is about allergic disorders, diseases in which dysregulated immune responses to ordinary environmental substances lead to intense, sometimes life-threatening, reactions. These include common diseases such as asthma, atopic dermatitis (eczema), and allergic rhinitis (hay fever), as well as rare disorders, including hypereosinophilic syndromes and vasculitides. The book is directed toward undergraduate and graduate students in immunology, as well as medical students and more advanced trainees in internal medicine and pediatrics. We assume that some readers will already have had exposure to a basic immunology course, whereas others will be naive to the basic principles of the field.

A case-based approach is used for several reasons. First is the insight often shared in speeches of graduating medical students that "patients are our best teachers." When we learn about the manifestations of a disease and its basic mechanisms in the context of a true human experience, the subject matter takes on a personal connection and really gels in our minds. Cases naturally catch our attention. Like a good mystery, the case-based process of diagnostic discovery draws the reader in. Finally, the work-up of unusual presentations of immune diseases reflects how immune mechanisms are discovered in reality. In this sense, patients are not only our best teachers but also represent the best experiments (of nature) that lead us to discovery. By working through these cases, our readers will therefore acquire an understanding of disease mechanisms in much the same way as the clinicians and researchers who initially described the diseases.

All of the cases relate to hypersensitivity reactions in which greatly enhanced sensitivity to normally innocuous substances leads to physiologic responses and tissue damage. In beginning immunology courses, students are often introduced to the immune system as a host defense center, the mechanism by which we discriminate self from nonself and eliminate pathogenic invaders. Actually, the immune system has an even more daunting job than just identifying and targeting germs. Among the outsiders to which our bodies are exposed every day, by inhalation, contact, and ingestion, the immune system must also discriminate those that are potentially harmful pathogens from those that are beneficial substances, such as foods, or harmless constituents of the environment, such as pollens and animal danders. The failure to either make that distinction or to actively suppress responses to those substances leads to allergy.

Allergic reactions have been around for most of recorded history; a description of the anaphylactic death from a wasp sting of the first Pharaoh of Egypt, King Menes, dates back to 2641 BC. Recent decades have seen an explosion in the incidence of allergies, particularly in developed nations. Allergic diseases now represent a major part of the practice of primary care providers. Food allergies, which were once rare, now affect more than 1 in 20 children in the United States, typically two students in every classroom. The familial inheritance pattern of allergies along with their rapidly increasing prevalence tells us that both genetic and environmental factors must drive these diseases. The cases that follow discuss both the known genetic factors (such as activating mutations of protein tyrosine kinases in mastocytosis and hypereosinophilic syndromes) and nongenetic contributors (including the bacteria colonizing our bodies and their immunomodulatory effects). Each case is used as a springboard for the description of important immune mechanisms, including mast cell and eosinophil development and function, mediators of hypersensitivity, mechanisms of immune regulation and tolerance, delayed-type hypersensitivity reactions, and complement function in angioedema.

Each of the 20 cases presented in this book derives from a real clinical scenario. Most of them describe patients seen in the clinical programs of Boston Children's Hospital. Names and places have been altered to maintain the privacy of our patients, but other details have been faithfully reproduced. Each case is presented in the same format. An introduction precedes the case, providing the basic scientific concepts required to understand the material. The case itself is presented as it would normally unfold in real life, beginning with a history of the patient's symptoms followed by a physical examination and a description of diagnostic studies that were performed. Doctors' notes are included in the margins to help illustrate the decision-making processes in diagnosis and treatment. After the case, the relevant disease process is summarized. Each case ends with a series of questions intended not as a quiz on the material presented but as a stimulus for further thinking about the disease mechanisms and their relationship to basic immunological principles.

Many colleagues have helped us in the preparation of this book. We are especially thankful to the pediatric residents and immunology fellows at Children's Hospital who have participated in the evaluations of these patients and provided clinical care. We thank Drs. A. Dvorak, S. Gellis, M. Lee, S. Vargas, T. Chatila, K. Willms, J. Orrell, G. Tsokos, V. Kyttaris, E.A. Morgan, and E.R. Tovey for contributed photomicrographs and patient images, and Drs. J. Boyce, O. Burton, and S. Leisten for helpful discussions. We are very grateful to the Garland Science team, including Janet Foltin, Allie Bochicchio, Eleanor Lawrence, and Ioana Moldovan, for their tireless work in assembling this book.

A note to the reader

The main topics addressed in each case correspond as much as possible to topics that are presented in the eighth edition of *Janeway's Immunobiology* by Kenneth Murphy. To indicate which sections of *Immunobiology* contain material relevant to each case, we have listed on the first page of each case the topics covered in it. The color code follows the code used for the five main sections of *Immunobiology*: yellow for the introductory chapter and innate immunity, blue for the sections on recognition of antigen, pink for the development of lymphocytes, green for the adaptive immune response, purple for the response to infection and clinical topics, and orange for methods.

Instructor Resources Website

Accessible from www.garlandscience.com, the Instructor Site requires registration and access is available only to qualified instructors. To access the Instructor Site, please contact your local sales representative or email science@garland.com.

The images from Case Studies in Allergic Disorders are available on the Instructor Site in two convenient formats: PowerPoint® and JPEG. They have been optimized for display on a computer. The resources may be browsed by individual cases and there is a search engine. Figures are searchable by figure number, figure name, or by keywords used in the figure legend from the book.

Contributors

Cem Akin, PhD, MD, Division of Rheumatology, Immunology and Allergy, Brigham and Women's Hospital; Associate Professor of Medicine, Harvard Medical School

Lisa Bartnikas, MD, Physician in Medicine, Division of Immunology, Boston Children's Hospital; Instructor in Pediatrics, Harvard Medical School

Sachin N. Baxi, MD, Physician in Medicine, Division of Immunology, Boston Children's Hospital; Instructor in Pediatrics, Harvard Medical School

Arturo Borzutzky, MD, Clinical Fellow in Pediatrics, Division of Immunology, Boston Children's Hospital; Research Fellow, Pediatrics, Harvard Medical School

Janet Chou, MD, Physician in Medicine, Division of Immunology, Boston Children's Hospital; Instructor in Pediatrics, Harvard Medical School

Ari Fried, MD, Physician in Medicine, Division of Immunology, Boston Children's Hospital; Instructor in Pediatrics, Harvard Medical School

James Friedlander, MD, Clinical Fellow in Pediatrics, Division of Immunology, Boston Children's Hospital

Mona Hedayat, MD, Research Fellow in Pediatrics, Division of Immunology, Boston Children's Hospital

John Lee, MD, Physician in Medicine, Division of Immunology, Co-director of Eosinophilic Gastrointestinal Disease Program, Boston Children's Hospital; Instructor in Pediatrics, Harvard Medical School

Mindy Lo, MD, PhD, Physician in Medicine, Division of Immunology, Boston Children's Hospital; Instructor in Pediatrics, Harvard Medical School

Andrew MacGinnitie, MD, PhD, Associate Clinical Director, Division of Immunology, Boston Children's Hospital; Assistant Professor of Pediatrics, Harvard Medical School

John Manis, MD, Investigator, Division of Transfusion Medicine, Boston Children's Hospital; Assistant Professor, Department of Pathology, Harvard Medical School

Itai Pessach, MD, PhD, Clinical Fellow, Division of Immunology, Boston Children's Hospital; Research Fellow, Pediatrics, Harvard Medical School

Frank J. Twarog, MD, PhD, Senior Associate in Medicine, Division of Immunology, Boston Children's Hospital; Clinical Professor of Pediatrics, Harvard Medical School

Christina Yee, MD, Physician in Medicine, Division of Immunology, Boston Children's Hospital; Instructor in Pediatrics, Harvard Medical School

Michael Young, MD, Physician in Medicine, Division of Immunology, Boston Children's Hospital; Assistant Clinical Professor of Pediatrics, Harvard Medical School

Contents

CASE 1 Acute Systemic Anaphylaxis

A life-threatening immediate hypersensitivity reaction to peanuts.

Adaptive immune responses can be elicited by antigens that are not associated with infectious agents. Inappropriate immune responses to otherwise innocuous foreign antigens result in allergic or hypersensitivity reactions, and these unwanted responses can be serious. Allergic reactions occur when an already sensitized individual is reexposed to the same innocuous foreign substance, or allergen. The first exposure generates allergen-specific antibodies and/or T cells; reexposure to the same allergen, usually by the same route, leads to an allergic reaction.

Acute systemic anaphylaxis is a type I IgE-mediated hypersensitivity reaction (Fig. 1.1) that is rapid in onset and can cause death. There is typically involvement of at least two organ systems, including the skin, respiratory, gastrointestinal, cardiovascular, or central nervous systems. As with any type I hypersensitivity reaction, the first exposure to the allergen generates allergen-specific IgE antibodies, which become bound to Fc receptors (FcεRI) on the surface of mast cells. On repeat exposure to allergen, cross-linking of IgE bound to FcεRI on mast cells and basophils leads to degranulation, with the release of preformed mediators such as histamine and tryptase and the synthesis and release of other mediators such as prostaglandins and leukotrienes. Histamine is a major mediator of the immediate effects of anaphylaxis, causing multiple symptoms including increased permeability of blood vessels, which can cause life-threatening hypotension. The mast-cell mediators important for anaphylaxis, and the clinical consequences of their release, are illustrated in Fig. 1.2.

Allergens introduced systemically are most likely to cause a serious anaphylactic reaction through the activation of sensitized connective tissue mast cells. The disseminated effects on the circulation and on the respiratory system are the most dangerous, and localized swelling of the upper airway can cause suffocation. Ingested antigens cause a variety of symptoms through their action on mucosal mast cells.

Any protein allergen can provoke an anaphylactic reaction, but those that most commonly cause acute systemic anaphylaxis are foods, medications, and insect venoms (Fig. 1.3). Proteins in food, most commonly milk, soy beans, eggs, wheat, peanuts, tree nuts, and shellfish, can also cause systemic anaphylaxis. Contact with protein antigens found in latex, a common constituent of

Topics bearing on this case:
Class I hypersensitivity reactions
Allergic reactions to food
Mast-cell activation via IgE

This case was prepared by Raif Geha, MD, in collaboration with Lisa Bartnikas, MD.

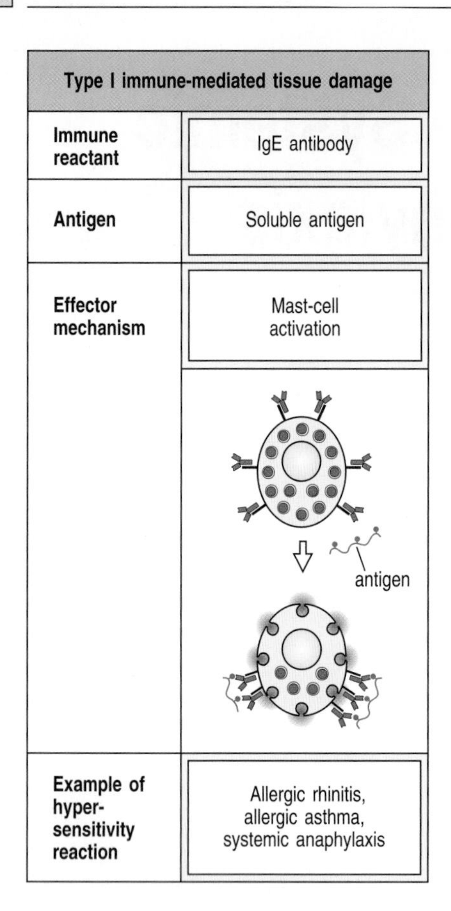

Type I immune-mediated tissue damage
Immune reactant
IgE antibody
Antigen
Soluble antigen
Effector mechanism
Mast-cell activation
antigen
Example of hyper-sensitivity reaction
Allergic rhinitis, allergic asthma, systemic anaphylaxis

Fig. 1.1 Type I immunological hypersensitivity reactions. Type I hypersensitivity reactions involve IgE antibodies and the activation of mast cells (see also Case 2).

rubber gloves, is also known to cause anaphylaxis. In addition, small-molecule antibiotics such as penicillin can act as haptens, binding to host proteins.

22-month-old child, unconscious, swollen face, difficulty in breathing. Give epinephrine immediately.

Type I allergic responses are characterized by the activation of allergen-specific CD4 helper cells (T_H2 cells) and the production of allergen-specific IgE antibody. The allergen is captured by B cells through their antigen-specific surface IgM and is processed so that its peptides are presented by MHC class II molecules to T-cell receptors of antigen-specific T_H2 cells. The interleukins IL-4 and/or IL-13 produced by the activated T_H2 cells induce a switch to the production of IgE, rather than IgG, by the B cell. However, allergen-specific IgE antibodies can exist without the occurrence of anaphylaxis, suggesting that factors other than IgE may be required.

This case concerns a child who suffered from life-threatening systemic anaphylaxis caused by an allergy to peanuts.

The case of John Mason: a life-threatening immune reaction.

John was healthy until the age of 22 months, when he developed swollen lips while eating cookies containing peanut butter. The symptoms disappeared in about an hour. A month later, while eating the same type of cookies, he started to vomit, became hoarse, had great difficulty in breathing, started to wheeze and developed a swollen face. He was taken immediately to the emergency room of the Children's Hospital, but on the way there he became lethargic and lost consciousness.

On arrival at hospital, his blood pressure was catastrophically low at 40/0 mmHg (normal 80/60 mmHg). His pulse was 185 beats min^{-1} (normal 80–90 beats min^{-1}), and his respiratory rate was 76 min^{-1} (normal 20 min^{-1}). His breathing was labored. An anaphylactic reaction was diagnosed and John was immediately given an intramuscular injection of 0.15 ml of a 1:1000 dilution of epinephrine (adrenaline). An intravenous solution of normal saline was infused as a bolus. The antihistamine Benadryl (diphenhydramine hydrochloride) and the anti-inflammatory corticosteroid Solu-Medrol (methylprednisolone) were also administered intravenously. A blood sample was taken to test for histamine and the enzyme tryptase.

Mediators of anaphylaxis		
Mediator	**Action**	**Signs/symptoms**
Histamine	Vasodilation, bronchoconstriction	Pruritus, swelling, hypotension, diarrhea, wheezing
Leukotrienes	Bronchoconstriction	Wheezing
Platelet-activating factor*	Bronchoconstriction, vasodilation	Wheezing, hypotension
Tryptase	Proteolysis	Unknown

Fig. 1.2 Mediators released by mast cells during anaphylaxis and their clinical consequences. *Platelet-activating factor is not released by mast cells but by neutrophils, basophils, platelets, and endothelial cells.

IgE-mediated allergic reactions			
Syndrome	**Common allergens**	**Route of entry**	**Response**
Systemic anaphylaxis	Drugs Serum Venoms	Intravenous (either directly or following oral absorption into the blood)	Edema Vasodilation Tracheal occlusion Circulatory collapse Death
Acute urticaria (wheal-and-flare)	Insect bites Allergy testing	Subcutaneous	Local increase in blood flow and vascular permeability
Allergic rhinitis (hay fever)	Pollens (ragweed, timothy, birch) Dust-mite feces	Inhaled	Edema of nasal mucosa Irritation of nasal mucosa
Allergic asthma	Danders (cat) Pollens Dust-mite feces	Inhaled	Bronchial constriction Increased mucus production Airway inflammation
Food allergy	Shellfish Milk Eggs Fish Wheat	Oral	Vomiting Diarrhea Pruritus itching Urticaria (hives) Anaphylaxis (rarely)

Fig. 1.3 IgE-mediated reactions to extrinsic antigens. All IgE- mediated responses involve mast-cell degranulation, but the symptoms experienced by the patient can be very different depending on whether the allergen is injected, inhaled, or eaten, and depending on the dose of the allergen.

Within minutes of the epinephrine injection, John's hoarseness improved, the wheezing diminished, and his breathing became less labored (Fig. 1.4). His blood pressure rose to 50/30 mmHg, the pulse decreased to 145 beats min^{-1} and his breathing to 61 min^{-1}. Thirty minutes later, the hoarseness and wheezing got worse again and his blood pressure dropped to 40/20 mmHg, his pulse increased to 170 beats min^{-1} and his respiratory rate to 70 min^{-1}.

John was given another intramuscular injection of epinephrine and was made to inhale nebulized albuterol (a β_2-adrenergic agent). This treatment was repeated once more after 30 minutes. One hour later, he was fully responsive, his blood pressure was 70/50 mmHg, his pulse was 116 beats min^{-1} and his respiratory rate had fallen to 46 min^{-1}. John was admitted to the hospital for further observation.

Treatment with Benadryl and methylprednisolone intravenously every 6 hours was continued for 24 hours, by which time the facial swelling had subsided and John's blood pressure, respiratory rate, and pulse were normal. He had stopped wheezing and when the doctor listened to his chest with a stethoscope it was clear.

He remained well and was discharged home with an EpiPen. His parents were instructed to avoid giving him foods containing peanuts in any form, and were asked to bring him to the allergy clinic for further tests.

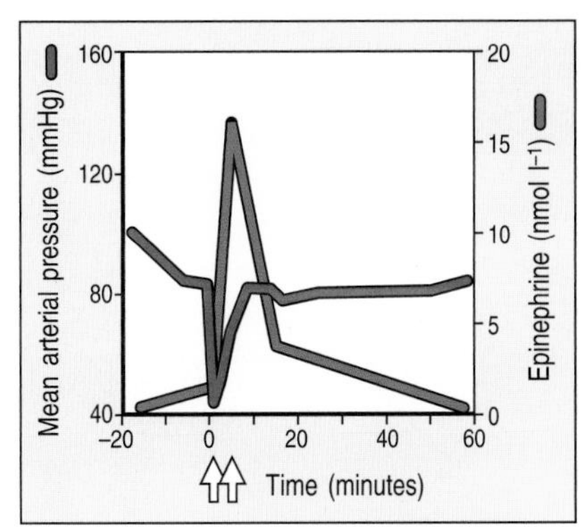

Fig. 1.4 Mean arterial pressure and epinephrine levels in a representative patient with insect-sting anaphylactic shock. Time 0 indicates the onset of the anaphylactic reaction as reported by the patient. The arrows indicate administration of antihistamines and epinephrine.

Acute systemic anaphylaxis.

Anaphylaxis presents a medical emergency and is the most urgent of clinical immunologic events; it requires immediate therapy. It results from the generation and release of a variety of potent biologically active mediators and their concerted effects on a number of target organs. John showed classic rapid-onset symptoms of anaphylaxis, starting with vomiting and swelling of the face and throat, and swelling of the wall of the bronchi along with constriction of the bronchial smooth muscle, which led to his difficulty in breathing. This was soon followed by a catastrophic loss of blood pressure, due to leakage of fluid from the blood vessels. Anaphylaxis can also cause urticaria (hives), heart arrhythmias, myocardial ischemia, and gastrointestinal symptoms such as nausea, vomiting, and diarrhea. All these signs and symptoms can occur singly or in combination.

Fatal allergic reactions to the venoms in bee and wasp stings have been recognized for at least 4500 years and account today for roughly 40 deaths each year in the United States. In 1902, Portier and Richet reported that a second injection of a protein from a sea anemone caused a fatal systemic reaction in dogs that had been injected previously with this protein. Because this form of immunity was fatal rather than protective, it was termed 'anaphylaxis' to distinguish it from the 'prophylaxis' (protection) generated by immunization.

Anaphylaxis requires a latent period for sensitization after the first introduction of antigen followed by reexposure to the sensitizing agent, which can be any foreign protein or a hapten. In the early part of the twentieth century, the most frequent cause of systemic anaphylaxis was horse serum, which was used as a source of antibodies to treat infectious diseases.

In many cases, the presentation of food allergy occurs on the first known ingestion, suggesting that routes other than the oral one may be important in sensitization. For example, epidemiologic data suggest that sensitization to peanut protein may occur in children through the application of peanut oil to inflamed skin. A recent study demonstrated that the incidence of peanut allergy in children who avoided peanut ingestion correlated with the level of peanut consumption in their homes, which is consistent with the skin being an important route of allergen sensitization. At present there is no cure for food allergy. Current therapy relies on allergen avoidance and the treatment of severe reactions with epinephrine.

Anaphylaxis is increasing in prevalence and is a frequent cause of visits to the emergency room, with 50–2,000 episodes per 100,000 persons, or a lifetime prevalence of 0.05–2.0%. The rate of fatal anaphylaxis from any cause is estimated at 0.4 cases per million individuals per year. Although in John's case the reaction was brought on by eating a food, an antigen administered by subcutaneous, intramuscular, or intravenous injection is more likely to induce a clinical anaphylactic reaction than one that enters by the oral or respiratory route.

Questions.

1 *Anaphylaxis results in the release of a variety of chemical mediators from mast cells, such as histamine and leukotrienes. Angioedema (localized swelling caused by an increase in vascular permeability and leakage of fluid into tissues) is one of the symptoms of anaphylaxis. With the above in mind, why did John get hoarse and why did he wheeze?*

2 *When his parents brought John back to the allergy clinic, a nurse performed several skin tests by pricking the epidermis of his forearm with a shallow plastic needle containing peanut antigens. John was also tested in a similar fashion with antigens from nuts as well as from eggs, milk, soy,*

and wheat. Within 5 minutes John developed a wheal, 10 mm × 12 mm in size, surrounded by a red flare, 25 mm × 30 mm (see Fig. 2.5), at the site of application of the peanut antigen. No reactions were noted to the other antigens. A fluorenzyme immunoassay (FEIA) was performed on a blood sample to examine for the presence of IgE antibodies against peanut antigens. It was positive. What would you advise John's parents to do?

3 *Why was John treated first with epinephrine in the emergency room?*

4 *Why was John given a blood test for histamine and the enzyme tryptase?*

5 *Why was the skin testing for peanuts not done in the hospital immediately after John had recovered, instead being done at a later visit?*

6 *The incidence of peanut allergy is increasing. Why?*

7 *John's parents want to know whether there are therapies that might cure him of his peanut allergy. What do you tell them?*

CASE 2 Allergic Asthma

Chronic allergic disease caused by an adaptive immune response to inhaled antigen.

Chronic allergic reactions are much more common than the acute systemic anaphylaxis reaction discussed in Case 1. Among these are allergic reactions to inhaled antigens (Fig. 2.1), which range in severity from a mild allergic rhinitis (hay fever) to potentially life-threatening allergic asthma, the disease discussed in this case. Once an individual has been sensitized, the allergic reaction becomes worse with each subsequent exposure to allergen, which not only produces allergic symptoms but also increases the levels of antibody and T cells reactive to the allergen.

Allergic asthma results from chronic T cell-driven inflammation of the airways with episodic symptoms triggered by allergen inhalation. The acute airflow obstruction experienced by an asthmatic patient following allergen exposure is an example of a type I hypersensitivity reaction. Type I reactions involve the activation of helper CD4 T_H2 cells, IgE antibody formation, mast-cell sensitization, and the release of allergic mediators. The allergen-specific IgE antibodies formed in sensitized individuals bind to and occupy high-affinity Fcε receptors (FcεRI) on the surfaces of tissue mast cells and basophils (Fig. 2.2). When the antigen is encountered again, it cross-links these bound IgE molecules, which triggers the immediate release of mast-cell granule contents, in particular histamine and various enzymes that increase blood flow and vascular permeability. This is the early phase of an immediate allergic reaction.

Within 12 hours of contact with antigen, a late-phase reaction occurs (Fig. 2.3). Arachidonic acid metabolism in the mast cell generates prostaglandins and leukotrienes, which further increase blood flow and vascular permeability. Cytokines such as interleukin-3 (IL-3), IL-4, IL-5, and tumor necrosis

Topics bearing on this case:
Inflammatory reactions
iNKT cells
Differential activation of T_H1 and T_H2 cells
IgE-mediated hypersensitivity
Skin tests for hypersensitivity
Radioimmunoassay
Tests for immune function

This case was prepared by Raif Geha, MD, in collaboration with Lisa Bartnikas, MD.

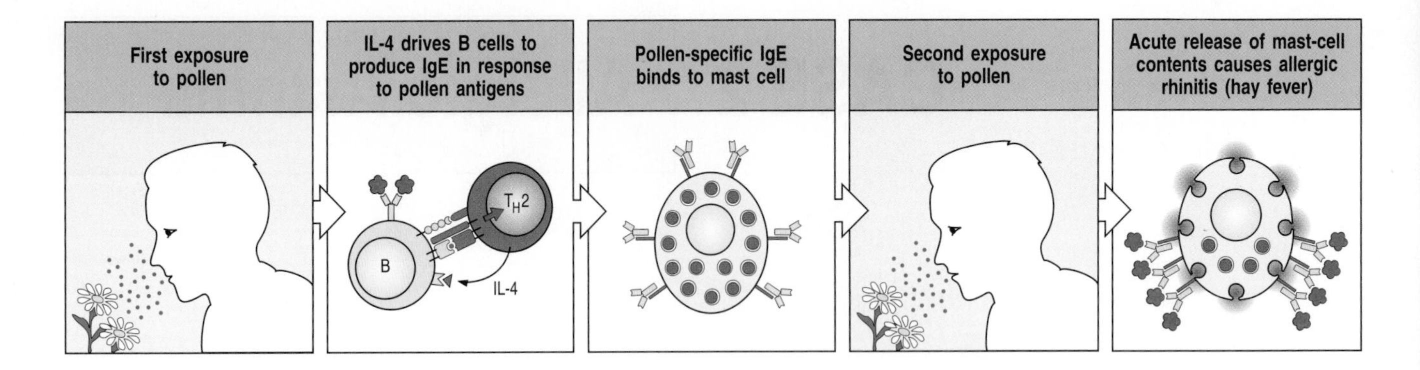

Fig. 2.1 Allergic reactions require previous exposure to the allergen. In this example, the first exposure to pollen induces the production of IgE anti-pollen antibodies, driven by the production of IL-4 by helper T cells (T_H2). The IgE binds to mast cells via FcεRI. Once enough IgE antibody is present on mast cells, exposure to the same pollen induces mast-cell activation and an acute allergic reaction, here allergic rhinitis (hay fever). Allergic reactions require an initial sensitization to the antigen (allergen), and several exposures may be needed before the allergic reaction is initiated.

factor-α (TNF-α) are produced by both activated mast cells and helper T cells, and these further prolong the allergic reaction. The mediators and cytokines released by mast cells and helper T cells cause an influx of monocytes, more T cells, and eosinophils into the site of allergen entry. The late-phase reaction is dominated by this cellular infiltrate. In the setting of recurrent allergen exposure, these cells infiltrating the allergic mucosa expand, particularly the eosinophils, and make a variety of products that are thought to be responsible for much of the tissue damage and mucus production that is associated with chronic allergic reactions. Cytokine-producing NKT cells have also been implicated in allergic asthma.

Approximately 15% of the population suffers from IgE-mediated allergic diseases. Many common allergies are caused by inhaled particles containing

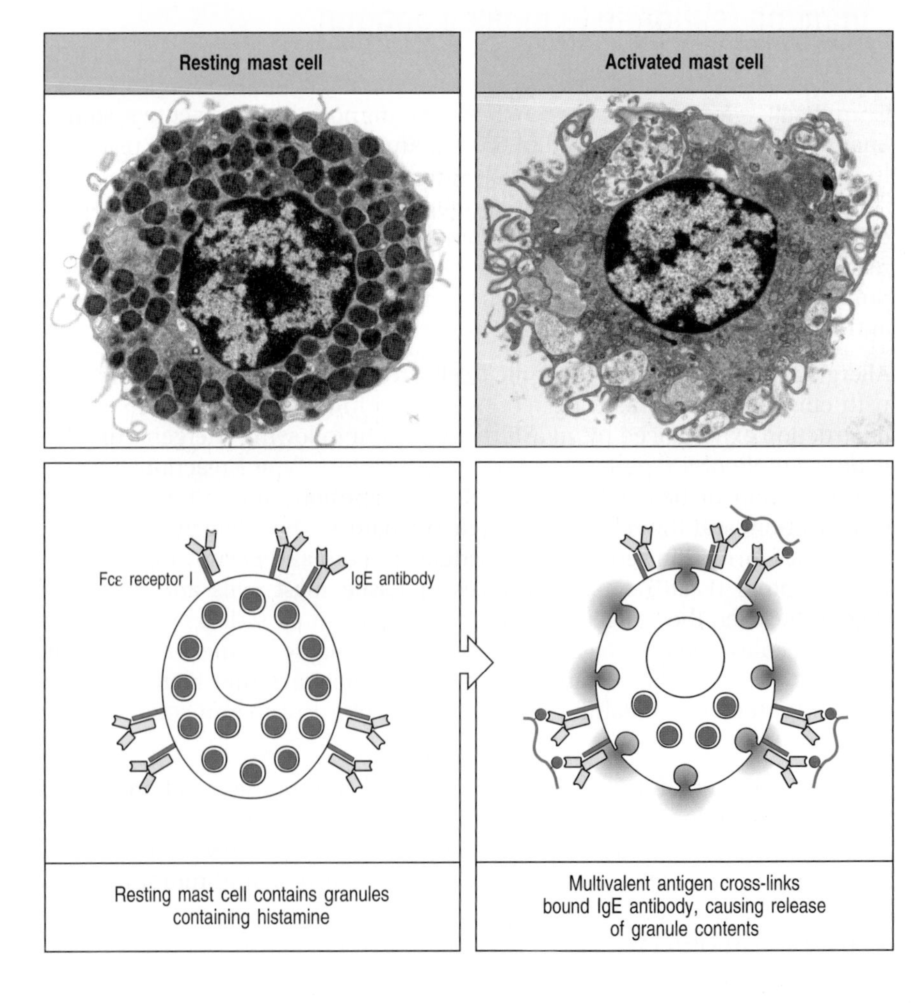

Fig. 2.2 Cross-linking of IgE antibody on mast-cell surfaces leads to a rapid release of inflammatory mediators by the mast cells. Mast cells are large cells found in connective tissue that can be distinguished by secretory granules containing many inflammatory mediators. They bind stably to monomeric IgE antibodies through the very high-affinity Fcε receptor (FcεRI). Antigen cross-linking of the bound IgE antibody molecules triggers rapid degranulation, releasing inflammatory mediators into the surrounding tissue. These mediators trigger local inflammation, which recruits cells and proteins required for host defense to sites of infection. It is also the basis of the acute allergic reaction causing allergic asthma, allergic rhinitis, and the life-threatening response known as systemic anaphylaxis (see Case 1). Photographs courtesy of A.M. Dvorak.

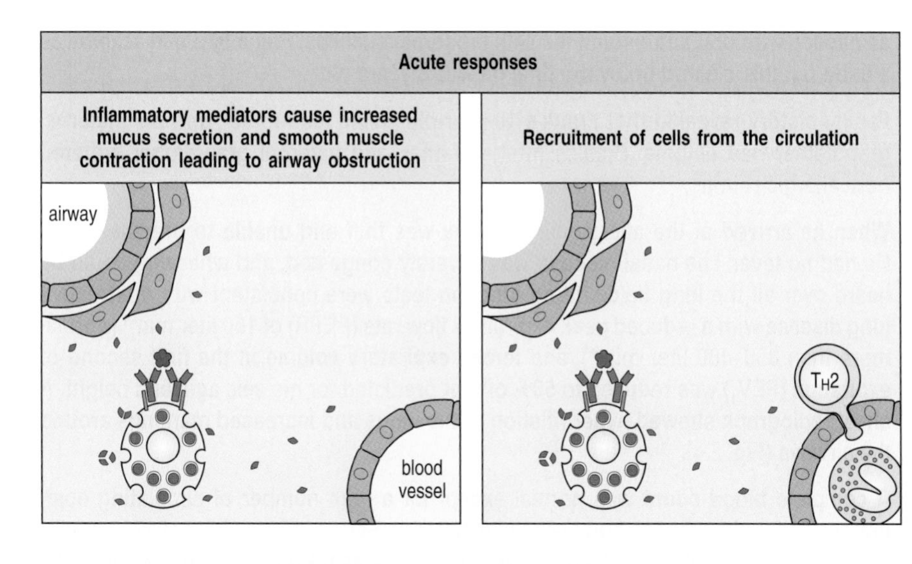

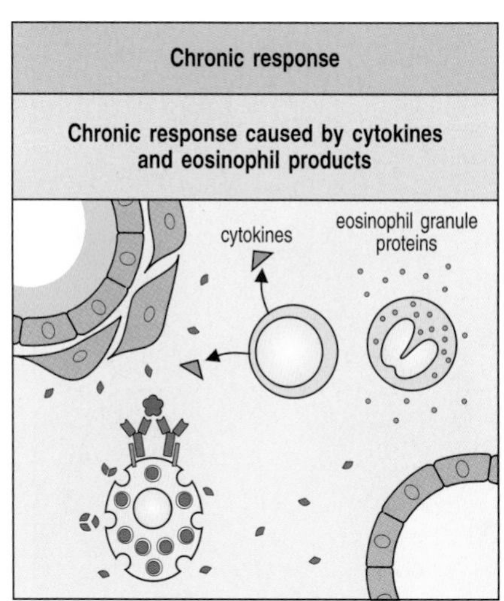

foreign proteins (or allergens) and result in allergic rhinitis, asthma, and allergic conjunctivitis. In asthma, the allergic inflammatory response increases the hypersensitivity of the airway not only to allergen reexposure but also to nonspecific agents such as exercise, pollutants, and cold air.

Fig. 2.3 The acute response in allergic asthma leads to T_H2-mediated chronic inflammation of the airways. In sensitized individuals, cross-linking of specific IgE on the surface of mast cells by inhaled allergen triggers them to secrete inflammatory mediators, causing bronchial smooth muscle contraction and an influx of inflammatory cells, including eosinophils and T_H2 lymphocytes. Activated mast cells and T_H2 cells secrete cytokines that also augment eosinophil activation, which causes further tissue injury and influx of inflammatory cells. The end result is chronic inflammation, which may then cause irreversible damage to the airways.

The case of Frank Morgan: a 14-year-old boy with chronic asthma and rhinitis.

Frank Morgan was referred by his pediatrician to the allergy clinic at 14 years of age because of persistent wheezing for 2 weeks. His symptoms had not responded to frequent inhalation treatment (every 2–3 hours) with a bronchodilator, the β_2-adrenergic agonist albuterol.

This was not the first time that Frank had experienced respiratory problems. His first attack of wheezing occurred when he was 3 years old, after a visit to his grandparents who had recently acquired a dog. He had similar attacks of varying severity on subsequent visits to his grandparents. Beginning at age 4 years, he had attacks of coughing and wheezing every spring (April and May) and toward the end of the summer (second half of August and September). A sweat test at age 5 years to rule out cystic fibrosis, a possible cause of chronic respiratory problems, was within the normal range.

As Frank got older, gym classes, basketball, and soccer games, and just going outside during the cold winter months could bring on coughing and sometimes wheezing. He had been able to avoid wheezing induced by exercise by inhaling albuterol 15–20 minutes before exercise. Frank had frequently suffered from a night-time cough, and his colds had often been complicated by wheezing.

Frank's chest symptoms had been treated as needed with inhaled albuterol. During the previous 10 years, Frank had been admitted to hospital three times for treatment of his asthma with inhaled bronchodilators and intravenous steroids. He had also been to the Emergency Room many times with severe asthma attacks. He had maxillary sinusitis at least three times, and each episode was associated with green nasal discharge and exacerbation of his asthma.

Since he was 4 years old, Frank had also suffered from intermittent sneezing, nasal itching, and nasal congestion (rhinitis), which always worsened on exposure to cats and dogs and in the spring and late summer. The nasal symptoms had been treated

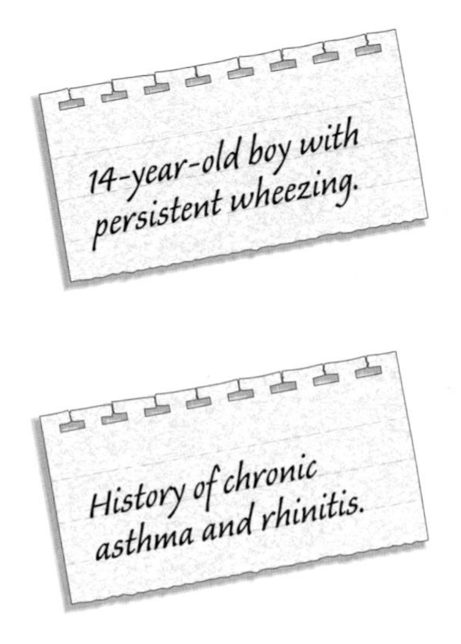

14-year-old boy with persistent wheezing.
History of chronic asthma and rhinitis.

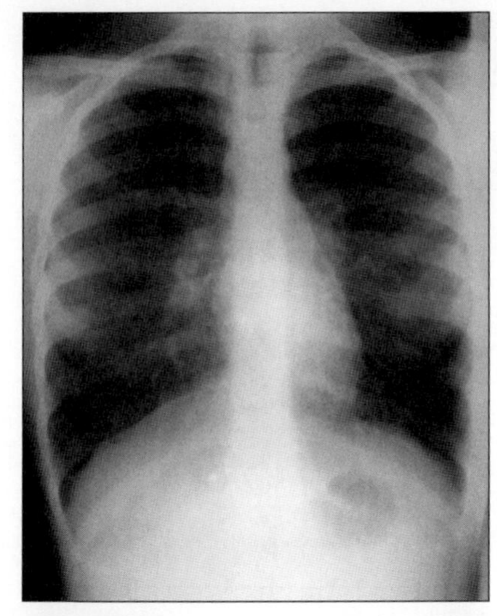

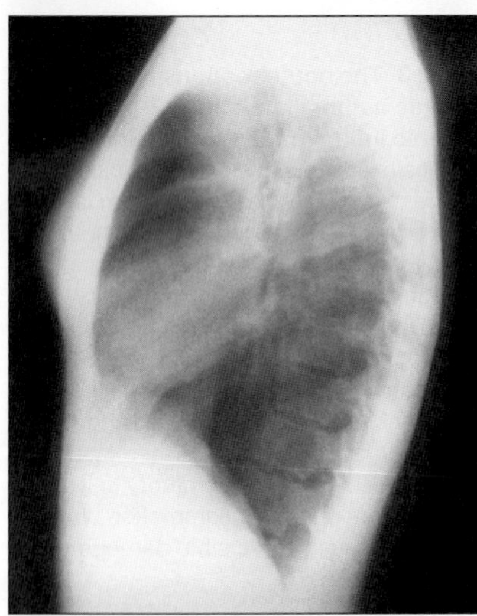

Fig. 2.4 Chest radiographs of a patient with asthma. Top: anteroposterior (A–P) view. Bottom: lateral view. The volume occupied by the lungs spans eight to nine rib spaces instead of the normal seven in the A–P view and indicates hyperinflation. The lateral view shows an increased A–P dimension, also reflecting hyperinflation. Hyperinflation indicates air trapping, which is a feature of the obstructive physiology seen in asthma. The bronchial markings are accentuated and can be seen to extend beyond one-third of the lung fields. This indicates inflammation of the airways.

as needed with oral antihistamines with moderate success. Frank had had eczema as a baby, but this cleared up by the time he was 5 years old.

Family history revealed that Frank's 10-year-old sister, his mother, and his maternal grandfather had asthma. Frank's mother, father, and paternal grandfather suffered from allergic rhinitis.

When he arrived at the allergy clinic, Frank was thin and unable to breathe easily. He had no fever. The nasal mucosa was severely congested, and wheezing could be heard over all the lung fields. Lung function tests were consistent with obstructive lung disease with a reduced peak expiratory flow rate (PEFR) of 180 liter min^{-1} (normal more than 350–400 liter min^{-1}), and forced expiratory volume in the first second of expiration (FEV_1) was reduced to 50% of that predicted for his sex, age, and height. A chest radiograph showed hyperinflation of the lungs and increased markings around the airways (Fig. 2.4).

A complete blood count was normal except for a high number of circulating eosinophils (1200 μl^{-1}; normal range less than 400 μl^{-1}). Serum IgE was high at 1750 ng dl^{-1} (normal less than 200 ng dl^{-1}). Fluorenzyme immunoassays (FEIA) for antigen-specific IgE revealed IgE antibodies against dog and cat dander, dust mites, and tree, grass, and ragweed pollens in Frank's serum. Levels of immunoglobulins IgG, IgA, and IgM were normal. Histological examination of Frank's nasal fluid showed the presence of eosinophils.

Frank was promptly given albuterol nebulizer treatment in the clinic, after which he felt better, his PEFR rose to 400 liter min^{-1}, and his FEV_1 rose to 65% of predicted. He was sent home on a 1-week course of the oral corticosteroid prednisone. He was told to inhale albuterol every 4 hours for the next 2–3 days, and then to resume taking albuterol every 4–6 hours as needed for chest tightness or wheezing. He was also started on fluticasone propionate (Flovent), an inhaled corticosteroid, and montelukast (Singulair), a leukotriene receptor antagonist for long-term control of his asthma. To relieve his nasal congestion, Frank was given the steroid fluticasone furoate (Flonase) to inhale through the nose, and was advised to use an oral antihistamine as needed. He was asked to return to the clinic 2 weeks later for follow-up, and for immediate hypersensitivity skin tests to try to detect which antigens he was allergic to (Fig. 2.5).

On the next visit Frank had no symptoms except for a continually stuffy nose. His PEFR and FEV_1 were normal. Skin tests for type I hypersensitivity were positive for multiple tree and grass pollens, dust mites, and dog and cat dander. He was advised to avoid contact with cats and dogs. To reduce his exposure to dust mites the pillows and mattresses in his room were covered with zippered covers. Rugs, stuffed toys, and books were removed from his bedroom. He was also started on immunotherapy with injections of grass, tree, and ragweed pollens, cat, dog, and house dust mite antigens, to try to reduce his sensitivity to these antigens.

A year and a half later, Frank's asthma continues to be stable with occasional use of albuterol during infections of the upper respiratory tract and in the spring. His rhinitis and nasal congestion now require much less medication.

Allergic asthma.

Like Frank, millions of adults and children suffer from allergic asthma. Asthma is the most common chronic inflammatory disorder of the airways and is characterized by reversible inflammation and obstruction of the small airways. Asthma has become an epidemic; the prevalence in the United States is increasing by 5% per year, with more than 500,000 new cases diagnosed

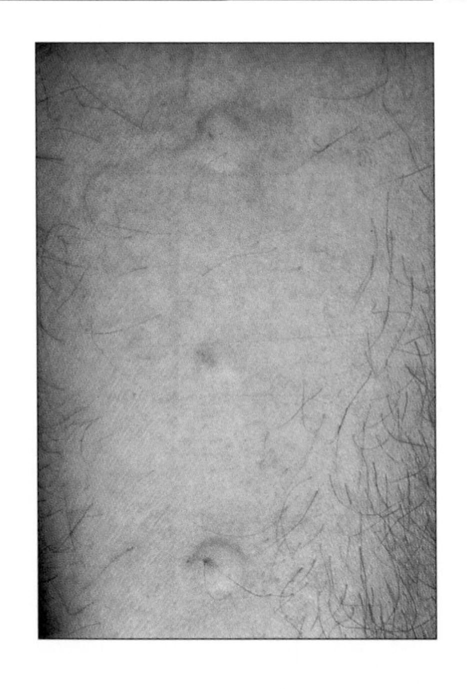

Fig. 2.5 An intradermal skin test. The photograph was taken 20 minutes after intradermal injections had been made with ragweed antigen (top), saline (middle), and histamine (bottom). A central wheal (raised swelling), reflecting increased vascular permeability, surrounded by a flare (red area), reflecting increased blood flow, is observed at the sites where the ragweed antigen and the positive histamine control were introduced. The small wheal at the site of saline injection is due to the volume of fluid injected into the dermis.

annually. It is the most common cause of hospitalization and days lost from school in children. About 70% of patients with asthma have a family history of allergy. This genetic predisposition to the development of allergic diseases is called atopy. Wheezing and coughing are the main symptoms of asthma, and both are due to the forced expiration of air through airways that have become temporarily narrowed by the constriction of smooth muscle as a result of the allergic reaction. As a consequence of the narrowed airways, air gets trapped in the lung, and the lung volume is increased during an attack of asthma (Fig. 2.6).

Once asthma is established, an asthma attack can be triggered not only by the allergen but by viral infection, cold air, exercise, or pollutants. This is due to a general hyperirritability or hyperresponsiveness of the airways, leading to constriction in response to nonspecific stimuli, thus reducing the air flow. The degree of hyperresponsiveness can be measured by determining the threshold dose of inhaled methacholine (a cholinergic agent) that results in a 20% reduction in airway flow. Airway irritability correlates positively with eosinophilia and serum IgE levels.

CD4 T cells are the central effector cells of airway inflammation in asthma. During asthma exacerbations, secretion of the T_H2-specific cytokines IL-4, IL-5, IL-9, and IL-13 is increased. Clinical improvement in asthma is associated with decreased T cells in the airways. Mast cells are also important effector cells in asthma and, after stimulation by allergen, release preformed

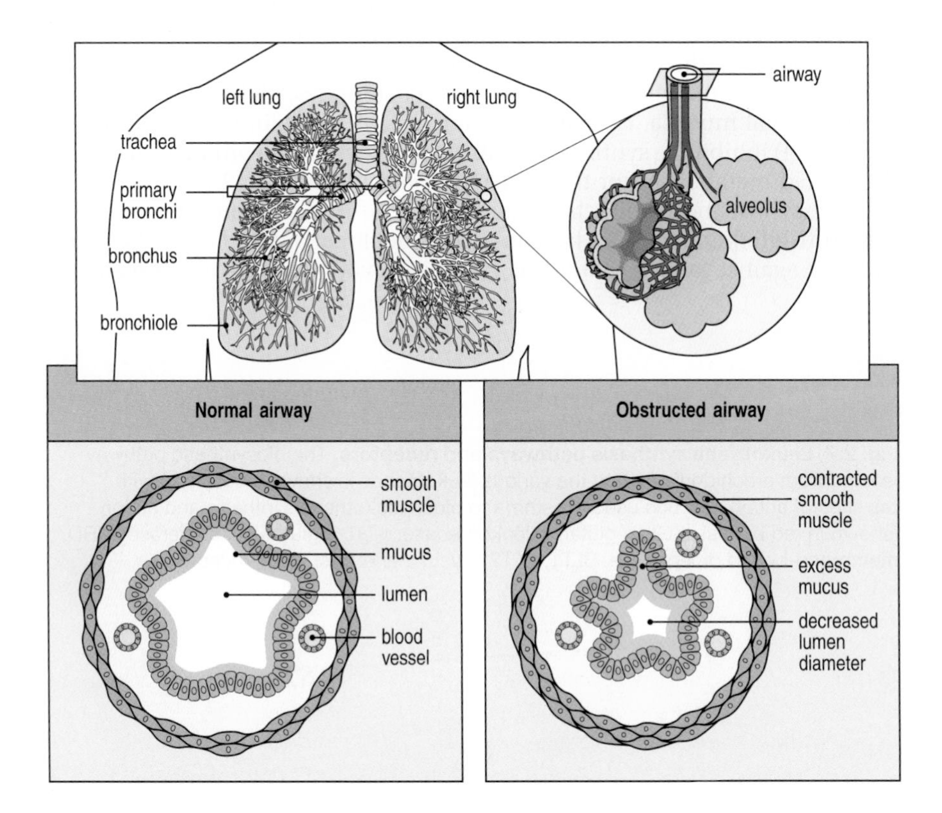

Fig. 2.6 Obstruction of the airways in chronic asthma. The top panels show the general anatomy of the lungs. Asthma is a chronic inflammatory disorder of the small airways—the bronchi and the bronchioles. In susceptible individuals, inflammation leads to recurrent wheezing, shortness of breath, chest tightness, and coughing. In between asthma attacks, patients are often asymptomatic, with normal physical exams and breathing tests. The bottom panels show schematic diagrams of sections through a normal airway (left) and an obstructed airway as a result of chronic asthma (right). During an asthma attack, there is infiltration of blood vessels of the small airways with immune cells (T_H2 lymphocytes and eosinophils), hypersecretion of mucus, and constriction and proliferation of bronchial smooth muscle. This leads to a decreased diameter of the airway lumen, resulting in wheezing and difficulty in breathing. In patients with severe asthma, there may be permanent airway remodeling.

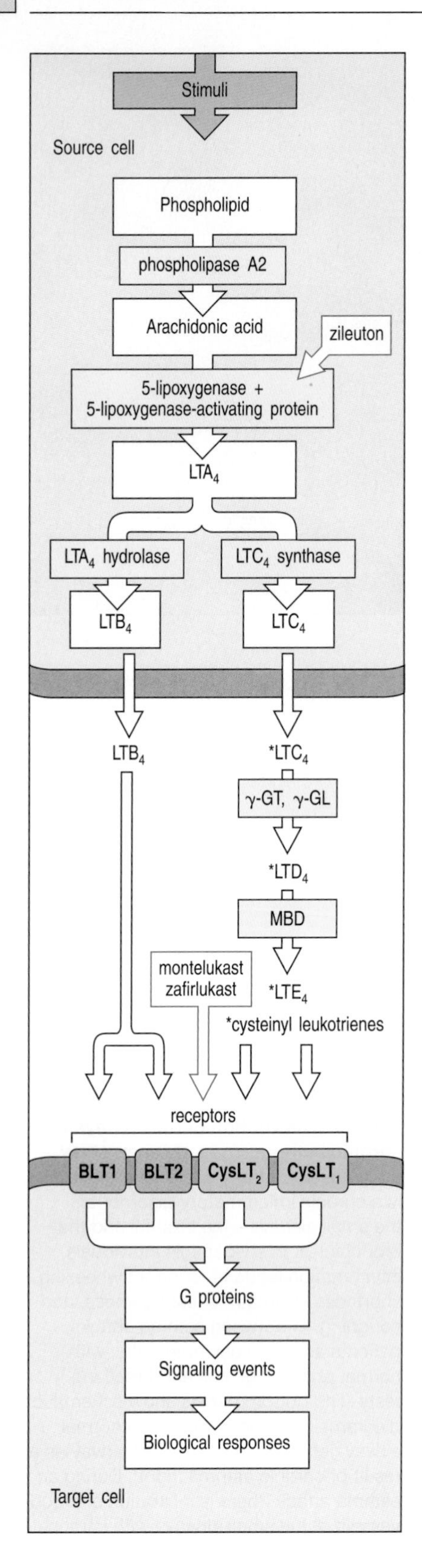

and newly generated mediators, contributing to acute and chronic mucosal inflammation. Cysteinyl leukotrienes, a product of arachidonic metabolism, are also key inflammatory mediators in asthma (Fig. 2.7). Cysteinyl leukotriene receptors include at least three types of transmembrane receptors. Activation of the cysteinyl leukotriene receptor 1 ($CysLT_1$) leads to bronchial smooth muscle constriction and muscle-cell proliferation, plasma leakage, hypersecretion of mucus, and eosinophil migration. The role of neutrophils in asthma is less clear. Elevated neutrophil numbers are more frequently seen in non-allergic asthma, steroid-unresponsive asthma, and in fatal asthma, suggesting that neutrophil-dominated asthma may represent a distinct asthma phenotype. Elevated neutrophil numbers in asthmatic lungs are associated with increased expression of IL-17.

The subset of T cells called invariant NKT cells (iNKT cells) is also elevated in asthmatic airways, suggesting that they may be important in human asthma. iNKT cells are a subpopulation of thymus-derived T cells that express markers of both T cells (such as the T-cell receptor:CD3 complex) and NK cells (such as NK1.1 and Ly-49). In humans, iNKT cells express an invariant antigen receptor with a variable region composed of $V_\alpha 24$–$J_\alpha 15$ paired with $V_\beta 11$. Unlike conventional T cells, iNKT cells can recognize glycolipid antigens bound and presented by the major histocompatibility complex (MHC) class Ib molecule CD1d. On activation, iNKT cells rapidly produce large amounts of the T_H1-type cytokine IFN-γ, the T_H2-type cytokines IL-4 and IL-13, and TNF-α and IL-2. The trigger for their activation in people with asthma could be glycolipids derived from microbes colonizing asthmatic airways.

Although asthma is a reversible disease, severe uncontrolled asthma can lead to airway remodeling, and a severe attack can be fatal. The mortality from asthma has been rising alarmingly in recent years. Risk factors for fatal asthma include frequent use of β_2-agonist therapy, poor perception of asthma severity, membership in a minority group, low socioeconomic status, adolescence, and male gender.

Several classes of drugs are commonly used to treat asthma, including corticosteroids, leukotriene antagonists, anti-IgE antibodies, anticholinergics, and β_2-adrenergic agonists. Corticosteroids (oral prednisone and inhaled fluticasone) inhibit the transcription of allergic and pro-inflammatory cytokines and can also activate the transcription of anti-inflammatory cytokines. This leads to a decrease in the numbers of mast cells, eosinophils, and T lymphocytes in the bronchial mucosa. Leukotriene antagonists (zileuton, montelukast, and zafirlukast) inhibit the synthesis of leukotrienes (which are products of arachidonic acid metabolism) or their receptor binding (see Fig. 2.7). Leukotriene modifiers have both mild bronchodilator and anti-inflammatory properties. Anti-IgE therapy uses a humanized monoclonal antibody (omalizumab) directed against the IgE that forms complexes with free IgE and prevents its

Fig. 2.7 Leukotriene synthesis pathways and receptors. The biosynthetic pathway leading from arachidonic acid to the various leukotrienes is shown here, along with the sites of action of drugs used in asthma to block leukotriene synthesis and action (shown in red boxes). γ-GL, γ-glutamyl leukotrienase; γ-GT, γ-glutamyl transferase; MBD, membrane-bound dipeptidase. BLT1, BLT2, $CysLT_2$, and $CysLT_1$ are receptors.

binding to the receptor FcεRI on the surfaces of mast cells and basophils. This results in a decrease in circulating free IgE and the downregulation of FcεRI expression on the cell surfaces. β_2-agonists (for example albuterol) bind to the β_2-adrenergic receptor, which is expressed on the surface of bronchial smooth muscle cells. β_2-agonists relax smooth muscle, thus rapidly relieving airway constriction, and are helpful in treating the immediate phase of the allergic reaction in the lungs. The treatment of allergic asthma also includes minimizing exposure to allergens and, in cases of severe or refractory environmental allergies, trying to desensitize the patient by immunotherapy (see Case 3).

Questions.

1 *Explain the basis of Frank's chest tightness and the radiograph findings.*

2 *Explain the failure of Frank's asthma to improve despite the frequent use of bronchodilators, and his response to steroid therapy.*

3 *Eosinophilia is often detected in the blood and in the nasal and bronchial secretions of patients with allergic rhinitis and asthma. What is the basis for this finding?*

4 *What is the basis of the wheal-and-flare reaction that appeared 20 minutes after Frank had had a skin test for hypersensitivity to ragweed pollen?*

5 *Frank called 24 hours after his skin test to report that redness and swelling had recurred at several of the skin test sites. Explain this observation.*

6 *Frank developed wheezing on several occasions after taking the nonsteroidal anti-inflammatory drugs (NSAIDs) aspirin and ibuprofen (Motrin). Explain the basis for these symptoms.*

7 *How would the immunotherapy that Frank received help to alleviate his allergies?*

8 *Although atopic children are repeatedly immunized with protein antigens such as tetanus toxoid, they almost never develop allergic reactions to these antigens. Explain.*

CASE 3 Allergic Rhinitis

A chronic stuffy, runny nose, cured by 'allergy shots.'

Every day we inhale myriad foreign proteins bound to small airborne particles. A subset of these, designated aeroallergens, has a propensity to elicit IgE antibody responses and, consequently, allergic reactions. Aeroallergens include proteins present in plant pollens and mold spores as well as epithelial, urinary, and salivary proteins associated with animal dander. Environmental levels of some aeroallergens, particularly the plant pollens, vary with the time of year, and individuals sensitized to them have corresponding seasonal flare-ups of symptoms. The molecular and immunological reasons why certain substances are allergenic are incompletely understood, but such substances all share the ability to induce specific IgE antibody responses in atopic individuals (that is, those with a genetic propensity to develop IgE-mediated hypersensitivity). Allergens are typically small water-soluble glycoproteins, and some have intrinsic protease activity. Many factors in addition to the allergen itself can influence the immune response to aeroallergens: they include concurrent viral infection, the profile of the microflora colonizing the host airway, and the presence of environmental pollutants.

The first contact between inhaled environmental allergens and the immune system occurs in the nasal mucosa. There, antigen-presenting cells sample aeroallergens and transport them to regional lymphoid tissues, where they are presented to CD4 helper T cells as peptide fragments in association with MHC class II molecules. These T cells are induced to differentiate into T_H2 effector cells, in which activation of the antigen receptor induces the expression of a pro-allergic constellation of cytokines, including interleukin-4 (IL-4), IL-5, and IL-13. T_H2 cells and their secreted cytokines are crucial in inducing B cells to produce IgE antibodies. This stimulation also requires physical interaction between the T cells and the B cells, which can occur either in the lymph nodes or directly in the nasal mucosa and is mediated by the binding of CD40 ligand on activated T cells to the CD40 receptor on B cells (Fig. 3.1).

A key step in the induction of IgE-mediated allergic responses is immunoglobulin class switching, in which antigen-activated B cells undergo a DNA recombination event that changes the class of immunoglobulin that they express. Class-switch recombination to IgE occurs in B cells bearing either IgM or IgG antigen receptors and which have been stimulated by two T_H2-cell-derived signals: IL-4, which activates the IL-4 receptor; and the activation antigen CD40 ligand (CD40L, also called CD154), which engages the TNF-receptor family member CD40 on the surface of the B cell. In combination, these signals activate a program of DNA transcription, induction of double-strand DNA breaks, and finally DNA repair and recombination, which result in the juxtaposition of

Topics bearing on this case:
Immune regulation
Immunoglobulin class switching
IgE antibodies
Mast-cell mediators

This case was prepared by Hans Oettgen, MD, PhD, and Raif Geha, MD, in collaboration with Sachin N. Baxi, MD.

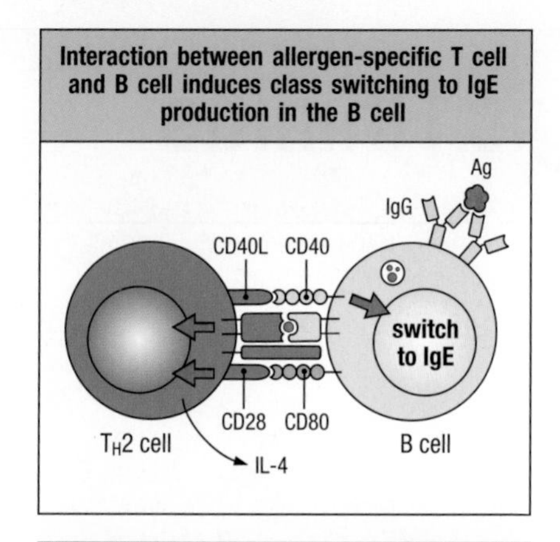
Interaction between allergen-specific T cell and B cell induces class switching to IgE production in the B cell
Ag
IgG
CD40L CD40
switch to IgE
CD28 CD80
TH2 cell
IL-4
B cell

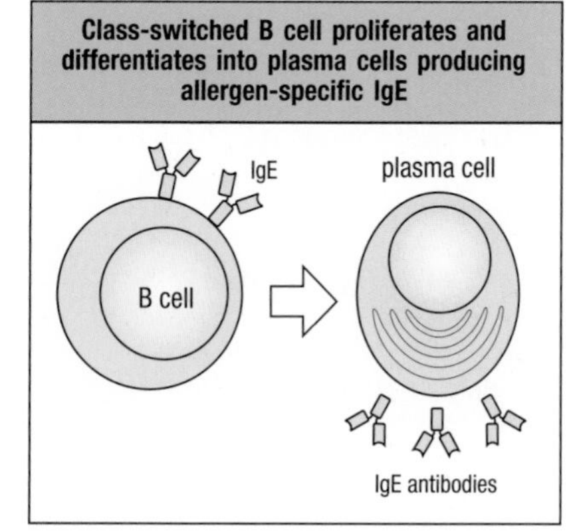
Class-switched B cell proliferates and differentiates into plasma cells producing allergen-specific IgE
IgE
plasma cell
B cell
IgE antibodies

Fig. 3.1 T-cell–B-cell interactions driving IgE switching. IgE production occurs after 'cognate' interaction between B cells and T_H2 cells sharing specificity for the same allergen. An allergen-specific IgM- or IgG-positive B cell binds allergen by means of its antigen receptor, internalizes it, and processes it in endocytic vesicles (top panel). Peptide fragments of the antigen are presented by the B cell to a neighboring antigen-specific T_H2 cell in association with MHC class II. Upon peptide:MHC recognition, the T-cell receptor (TCR) induces signals that result in the transcription of cytokine genes and cytokine production (IL-4 in this T_H2 cell) and the rapid induction of the cell-surface protein CD40 ligand. CD40L binds CD40, which is constitutively present on B cells. The CD40 signal both induces activation of the IgE class-switch recombination machinery in the B cell's nucleus and the expression of co-stimulatory molecules, including CD80 family members, which bind to receptors (CD28) on the T cell, resulting in the amplification of IL-4 production. The combination of signals from IL-4 and CD40L induces class switching to IgE in the B cell, which subsequently differentiates into IgE- producing plasma cells (bottom panel).

the VDJ cassette encoding the rearranged immunoglobulin heavy-chain variable region and the DNA encoding the constant-region domains of the ε heavy chain (Fig. 3.2). Careful analysis for the presence of RNA transcripts and DNA splicing by-products (switch-excision circles) has shown that IgE switching occurs not only in lymph nodes but also directly in allergen-exposed mucosal tissues (Fig. 3.3).

In individuals producing allergen-specific IgE, re-exposure to the allergen leads to cross-linking of FcεRI-bound IgE on mast cells, resulting in mast-cell degranulation and the release of preformed allergic mediators (histamine, proteases, and proteoglycans) as well as newly synthesized mediators (prostaglandins, leukotrienes, and platelet-activating factor). These mediators elicit both the immediate reactions of swelling, redness, and itching (pruritis) of an allergic reaction as well as late-phase responses 8–12 hours later (see Case 4).

The case of Charlie Marlow: a 14-year-old boy with chronic rhinitis.

One spring, Charlie Marlow's pediatrician referred him to the allergy clinic at Children's Hospital for evaluation of a runny, stuffy, and itchy nose accompanied by red, itchy, watery eyes. Charlie was 14 years old at the time. The thing that bothered him the most was that he could no longer breathe through his nose. He had had these symptoms for 4 weeks, beginning in early spring, and over-the-counter cold remedies had not brought any relief. In fact, the side effects of the medicines had made it difficult for Charlie to concentrate, which combined with the nasal symptoms made him miserable. He recalled having had similar but milder symptoms the previous year, but told the allergist that for the most part his nose had been fine over the summer, fall (autumn) and winter months.

Charlie lived with his parents and sister in a 10-year-old single-family home with central air conditioning and hot water heating. There were no pets at home and no issues with mold, mildew, mice, or cockroaches. Nobody smoked in the home.

Charlie was continually sniffing and constantly rubbing his nose with the palm of his hand. He was obviously breathing through the mouth and had a nasal quality to his voice. His conjunctivae were slightly swollen. He had dark skin around his lower eyelids with the appearance of 'allergic shiners.' The tympanic membranes were normal. The nasal mucosa appeared pale and bluish in color and was markedly swollen. There was copious clear drainage and virtually no air movement through the nares. A crease was visible just below the bridge of the nose that had arisen from Charlie's constant

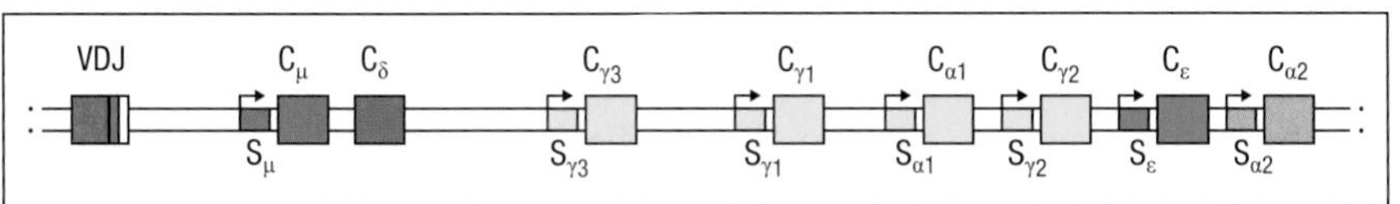

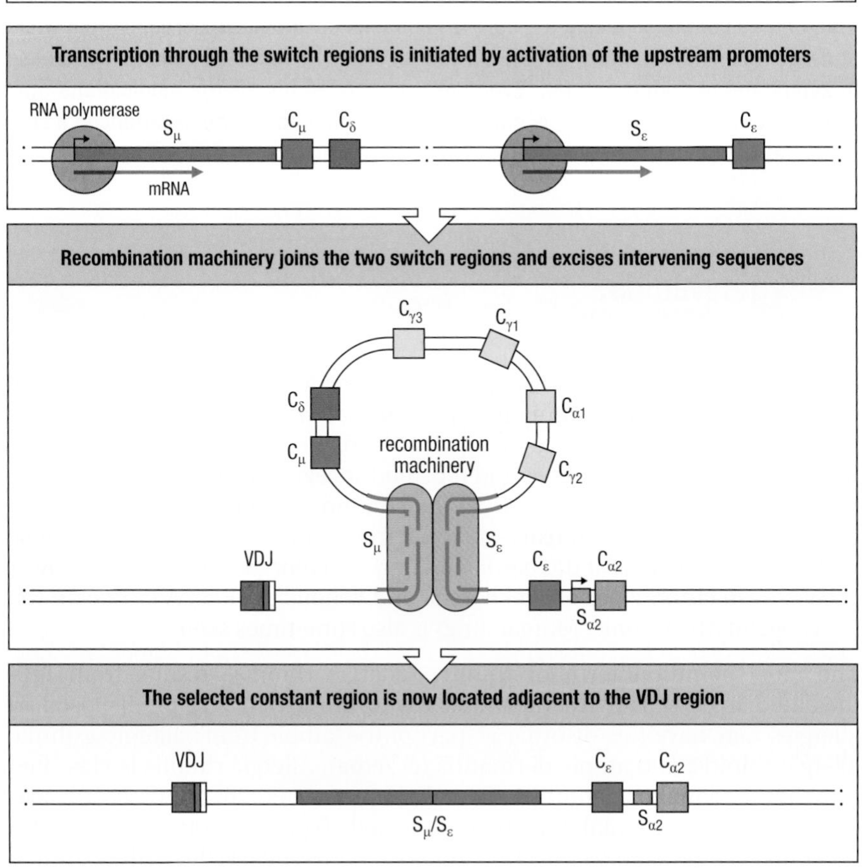

Fig. 3.2 Immunoglobulin class switching involves recombination between specific switch signals. The organization of a rearranged immunoglobulin heavy-chain locus before class switching (top panel). Switch regions (S), repetitive DNA sequences that guide class switching, are found upstream of each of the immunoglobulin C genes, with the exception of the δ gene. Switching is guided by the cytokine-induced initiation of transcription (shown as arrows) through these regions from promoters located upstream of each S. Switching between the μ and ε isotypes (second panel). Stimulation with IL-4 or IL-13 induces transcription at the IgE S region, resulting in the production of ε-germline transcripts (εGLT). CD40L ligation of CD40 and the activation of the IL-4 receptor on the B cell (see Fig. 3.1) induce the enzyme activation-induced cytidine deaminase (AID). AID is recruited to the sites of transcription in the immunoglobulin locus (not shown), where it activates class-switch recombination. The introduction of DNA double-strand breaks results in recombination between the targeted S regions (third panel). During recombination, the intervening genomic DNA between S_μ and S_ε is deleted as a switch-excision circle. Switch-excision circles are present only in newly switched B cells and are diluted upon cell division. Recombination generates chimeric S_μ/S_ε regions composed of the 5′ S_μ joined to the 3′ portion of the targeted S_ε region and brings the C_ε exons adjacent to the rearranged VDJ exon (bottom panel). Transcription through this region generates ε heavy-chain RNA.

pushing and rubbing on the nose. The remainder of the physical examination was normal for his age.

Microscopic examination of a nasal smear stained with hematoxylin and eosin revealed the presence of abundant eosinophils and no neutrophils, confirming the diagnosis of allergic rhinitis.

Skin testing was performed to determine whether Charlie had IgE-mediated sensitivity to environmental allergens. Small drops of aqueous extracts of several allergens were applied to the skin in a grid. The skin was then gently pricked at the site of each drop to allow the extracts to penetrate into the dermis. Charlie began to sense intense itching at some of the spots within minutes. Fifteen minutes after application, reactions were observed with tree pollens, grass pollen, molds, dog, cat, mouse, and dust mites. These reactions had a central wheal, indicative of dermal vascular leak and local edema, with a surrounding 'flare,' a halo of erythema induced by a local neurogenic increase in blood flow. This wheal-and-flare response is the cutaneous form of immediate hypersensitivity, and these reactions confirmed the presence of IgE antibodies specific for these allergens.

Allergies to: pollens, molds, dog, cat, dust mites.

Charlie's younger sister was disappointed to see that he was allergic to animals. She had wanted the family to get a cat but they were now advised against this. A treatment plan was designed based on allergen avoidance and the use of long-acting, non-sedating oral antihistamines (which do not cross the blood–brain barrier) and an intranasal corticosteroid spray.

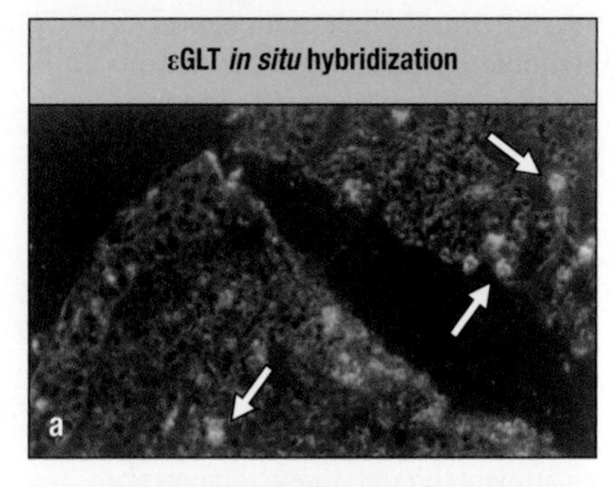

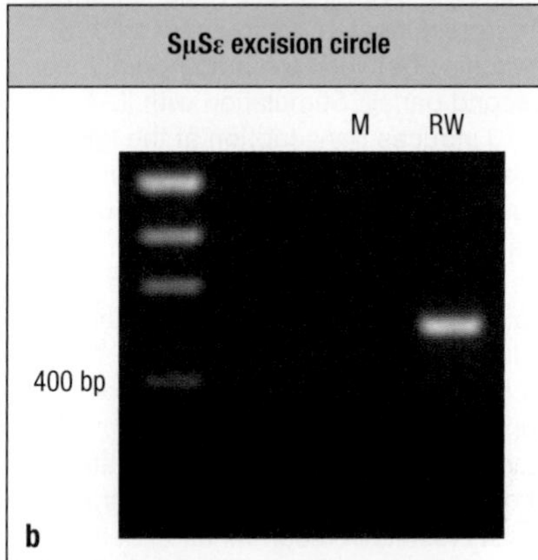

Fig. 3.3 The detection of IgE class-switch recombination in the nasal mucosa. The culture of nasal mucosal biopsies from ragweed-allergic subjects with ragweed allergen (RW) results in the production of εGLT RNA (panel a), which is detected by *in situ* hybridization, and of switch-excision circles (panel b; see Fig. 3.2), which are detected by PCR and electrophoresis. The band representing switch-excision circle DNA can be seen on the right in panel b; bands on the left are size markers. These results provide strong evidence that IgE switching is not restricted to secondary lymphoid organs but can occur in B cells in the nasal mucosa. Photographs courtesy of L. Cameron.

This plan worked very well for a few years but over time Charlie's symptoms increased in severity and duration and began to manifest all the year round, and his medications became less and less effective. He was offered the option of immunotherapy or 'allergy shots.' This involved monthly subcutaneous injections of allergen extracts. At first, Charlie continued taking the daily antihistamines and nasal corticosteroid while undergoing immunotherapy, but after a year of shots he reported a marked decrease in symptoms in spring and fall. Eventually he needed his allergy medications only rarely. After five years his allergy shots were discontinued and he experienced a sustained relief from his allergy symptoms even after that.

Allergic rhinitis.

Rhinitis is an inflammatory disease of the upper airway characterized by nasal congestion, rhinorrhea (runny nose), postnasal drip, sneezing, and an itchy nose. It can be accompanied by symptoms involving the eyes, ears, and throat. Itching of mucous membranes and repeated sneezing are the most characteristic complaints. Charlie showed all the commonly seen symptoms of allergic rhinitis, including a transnasal crease from continually rubbing the nose upward with the palm of the hand, and dark discoloration beneath the lower eyelids from chronic periorbital edema and venous pooling. A crease on the lower eyelid (the Dennie–Morgan line) is also sometimes seen.

The most common form of rhinitis—allergic rhinitis—results from IgE-mediated hypersensitivity. Individuals who are genetically predisposed to allergies may have this disorder as part of the 'atopic triad': allergic asthma, allergic rhinitis, and atopic dermatitis (eczema). Allergic rhinitis is classified as either seasonal or perennial. Overall, an estimated 20% of cases are seasonal, 40% are perennial, and 40% are mixed. Typical allergenic triggers for perennial allergic rhinitis include indoor allergens such as dust mites, animal dander, and rodents. Seasonal rhinitis is induced by outdoor allergens such as pollens, which typically drive seasonal symptoms related to the timing of pollen release by the various plants, including trees (spring), grasses (summer), and weeds (fall) that can trigger symptoms. Allergic rhinitis is estimated to affect 60 million people in the United States. It is the most common atopic disease, affecting approximately 40% of children and between 10% and 30% of adults.

In most parts of the United States, pollen is produced by trees in the spring, grass from late spring to early summer, and weeds from late summer through the fall (Fig. 3.4). In tropical climates, growing seasons are longer. Plant pollens are seasonally produced in very large amounts—a single birch flower cluster produces around 6 million pollen grains, which can be wafted on the breeze for many miles. Grass pollen is a common cause of spring and summer allergies throughout the United States. Pollen from weeds such as ragweed, lamb's quarter, and plantain is present throughout the United States.

Spores of molds such as *Cladosporium*, *Aspergillus*, and *Penicillium* can be responsible for both seasonal and perennial allergic symptoms. Spores of outdoor fungi peak in the mid-summer and diminish with the first hard frost in regions that have cold winters. *Alternaria* is the most prevalent outdoor mold in dry climates, and is found in soil, seeds, and plants.

Spores of *Cladosporium*, which grows on decaying plant matter, are the most prevalent fungal allergen in temperate regions. *Aspergillus* is often isolated from house dust and is found in compost heaps and dead vegetation. *Penicillium* is present in soil, grains, and house dust. It grows in water-damaged walls, giving a green 'mildew' color. Although it is not possible to control

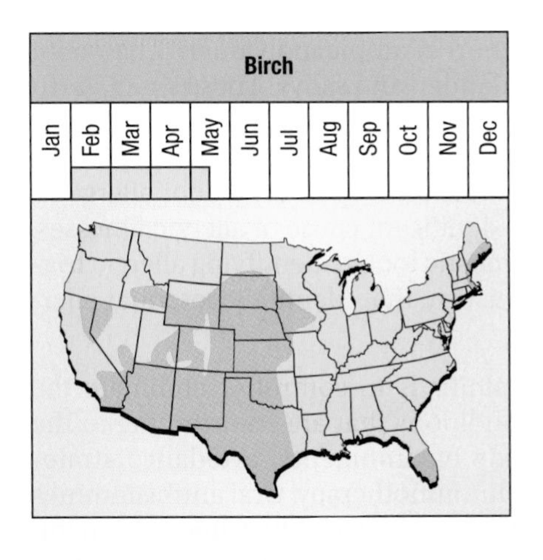

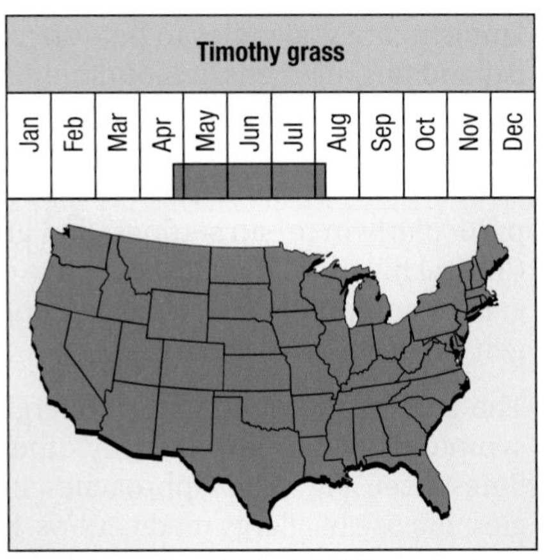

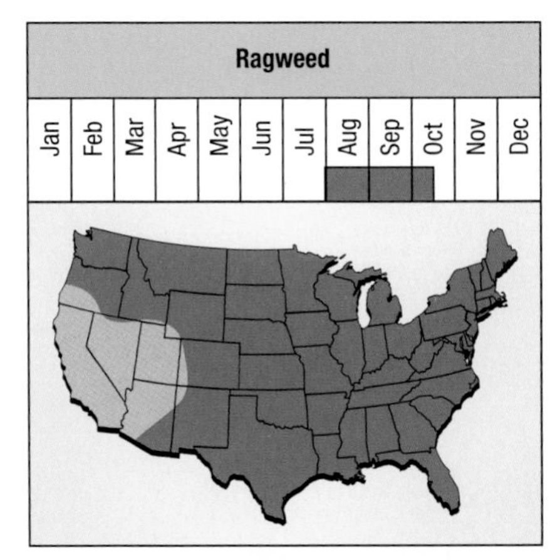

Fig. 3.4 Seasonal and distribution patterns for aeroallergens in the United States. Pollen is typically produced by trees in the spring, grass in the late spring to summer, and weeds in the fall. The maps represent the geographical distributions of some common pollens from these three classes; and the geographical areas illustrated in pink represent areas negative for the presence of pollen.

pollen or mold spore levels outside the home, one can reduce the amount of indoor allergen. Allergen-avoidance measures are described in Fig. 3.5.

Perennial allergic rhinitis is characterized by year-round nasal symptoms driven by allergens that do not vary with the season. Such allergens include house dust mites, pet dander, cockroaches, and mice. House dust mites (*Dermatophagoides*) are microscopic arachnids (Fig. 3.6) that are found in dust and in things made of woven materials such as mattresses, pillows, stuffed animals, and bedding. The allergenic proteins are contained in the mite feces. Several studies have demonstrated that exposure to dust mites in early childhood is an important determinant in both sensitization—the production of dust-mite-specific IgE antibodies—and in asthma development.

Allergen source	Avoidance measures
Dust mites	Encase the pillow and mattress Wash bedding weekly in hot water and dry in a heated dryer Remove reservoirs such as toys and stuffed animals from the bed Keep humidity at less than 50% Vacuum with a HEPA filter bag
Pets	Remove pet from the home If cannot remove: Remove pet from the bedroom Aggressively clean upholstered furniture, walls, and carpet Encase the mattress and pillows Keep the pet clean. Washing dogs twice a week may be helpful Use a HEPA air filter
Rodents	Detect hiding places and food sources (grease, kitchen debris) Store food in sealed containers Remove clutter Exterminate with pesticide or bait traps Seal holes or cracks in the home
Pollen	Keep windows closed Bathe to remove allergens from hair and body HEPA air filtration
Mold	Locate mold-contaminated items and remove them Clean moist areas as these are prone to mold growth Repair leaks Reduce humidity to less than 50% For outdoor mold, follow the same advice as for pollen

Fig. 3.5 Allergen avoidance measures.

Unfortunately, allergies to pets are common in atopic individuals. The major cat and dog allergens are found in hair, dander, and saliva. These small particles scatter easily in the air and adhere to clothing, which aids their wide dispersal. They can be detected in schools and other public places where pets are not normally present. Cockroaches and mice are important indoor allergens, particularly in urban settings, and are a significant cause of allergic illnesses. Clinical history is the most effective diagnostic tool in identifying allergic triggers. A comprehensive environmental history that identifies potential allergen exposures is essential.

The goal of the treatment of allergic rhinitis is to control or eliminate the symptoms and to improve any other conditions that are contributing to the illness (see Fig. 3.5). Approaches include environmental-avoidance strategies, the use of allergy medications, and immunotherapy. Oral antihistamines reduce itchiness, sneezing, and rhinorrhea, and are a first-line treatment. Antihistamines function as 'inverse agonists,' stabilizing the inactive form of the histamine H1 receptor (H1R), making it unable to bind histamine and so preventing it from triggering cellular responses. The main adverse side effects of the first generation of antihistamines were sedation and anticholinergic effects (dry mouth and eyes). Newer antihistamines are preferred because they have a longer duration of action and reduced sedative and anticholinergic effects. Intranasal corticosteroids are the most effective treatment for reducing seasonal and perennial allergic rhinitis. They reduce nasal congestion, sneezing, itching, rhinorrhea, and postnasal drip. Intranasal corticosteroids

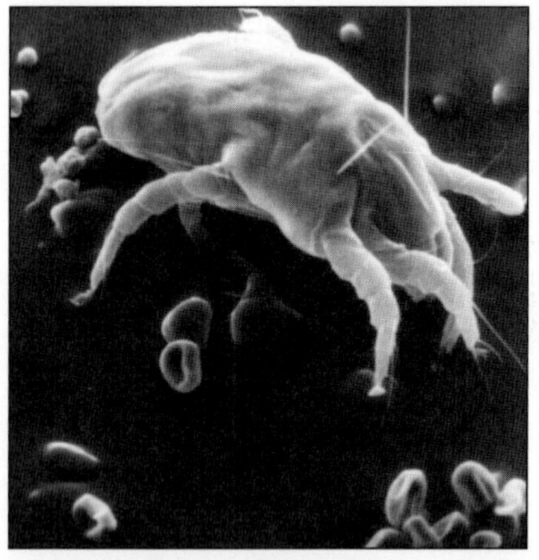

Fig. 3.6 Dust mite. The scanning electron micrograph in the upper panel shows a dust mite with fecal pellets. These microscopic arachnids are commonly found in woven materials such as pillows and mattresses. The course of events leading to sensitization to inhaled allergens is shown in the bottom panels. Der p 1, found in the fecal pellets of house dust mites, is a common perennial aeroallergen. On a first encounter of an atopic individual with Der p 1, T_H2 cells specific for Der p 1 may be produced (first and second panels). Interaction of these T cells with Der p 1-specific B cells leads to the production of class-switched plasma cells producing Der p 1-specific IgE in the mucosal tissues (third panel), and this IgE becomes bound to Fc receptors on resident submucosal mast cells. On a subsequent encounter with Der p 1, the allergen binds to the mast-cell bound IgE, triggering mast-cell activation and the release of mast-cell granule contents, which cause the symptoms of allergic rhinitis (last panel). Der p 1 is a protease that cleaves occludin, a protein that helps to maintain the tight junctions in epithelia; the enzymatic activity of Der p 1 is thought to help it pass through the epithelium. Photograph courtesy of E.R. Tovey.

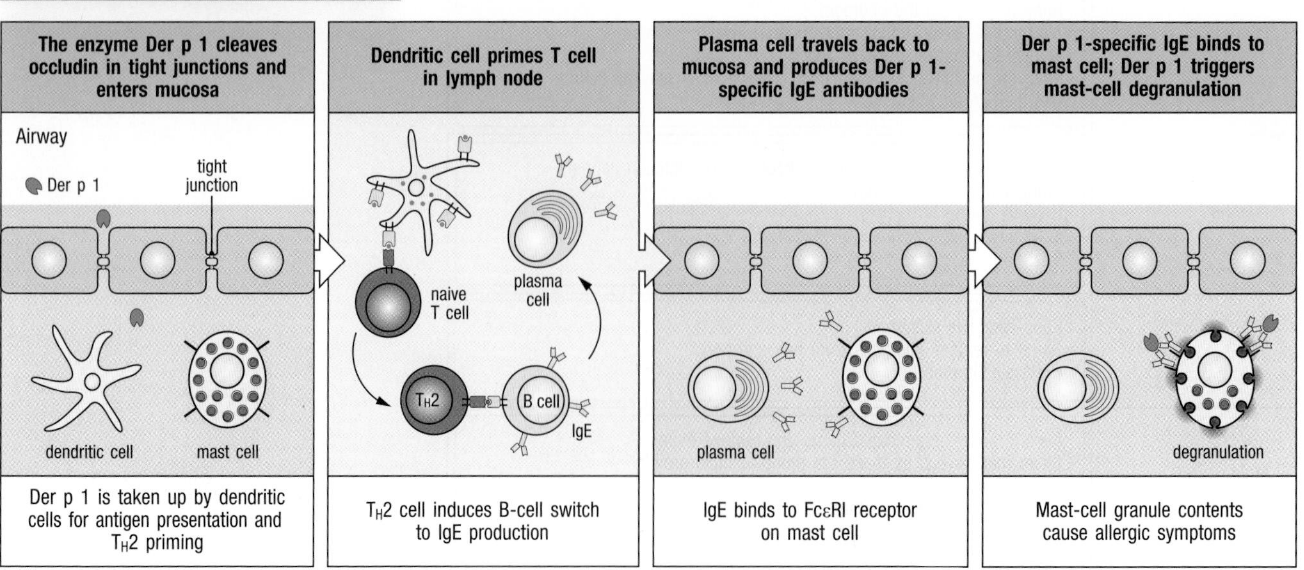
The enzyme Der p 1 cleaves occludin in tight junctions and enters mucosa
Airway
Der p 1
tight junction
dendritic cell
mast cell
Der p 1 is taken up by dendritic cells for antigen presentation and TH2 priming
Dendritic cell primes T cell in lymph node
naive T cell
plasma cell
TH2
B cell
IgE
TH2 cell induces B-cell switch to IgE production
Plasma cell travels back to mucosa and produces Der p 1-specific IgE antibodies
plasma cell
IgE binds to FcεRI receptor on mast cell
Der p 1-specific IgE binds to mast cell; Der p 1 triggers mast-cell degranulation
degranulation
Mast-cell granule contents cause allergic symptoms

work by suppressing inflammation through the downregulation of inflammatory mediators and cytokines, and start to act 3–12 hours after administration.

The concept of allergen immunotherapy or allergy shots was first introduced by Leonard Noon in 1911. He observed that periodic subcutaneous injection of increasing doses of grass-pollen extract over several months resulted in a reduction of symptoms in allergic subjects. The basic principles of subcutaneous immunotherapy (SCIT) remain the same today, but there is now a wide range of high-quality standardized extracts whose therapeutic doses are well characterized. Recently, sublingual immunotherapy, in which allergen doses 20–200 times those used for SCIT are applied under the tongue, has been introduced into clinical practice.

Immunotherapy has been shown to be especially effective in individuals with venom hypersensitivity and those with allergic rhinitis. There is evidence that the repeated injection of allergen has several immunological effects that act in concert to blunt allergic responses (Fig. 3.7). Typically, individuals with allergic rhinitis have IgE antibodies specific for the offending allergen and exhibit seasonal increases in allergen-specific IgE. Although immunotherapy does suppress baseline anti-allergen IgE levels, the seasonal increase is blunted and seems to be replaced by an increase in IgG4 antibodies specific for allergen. There is a good correlation between the induction of IgG4 and the clinical efficacy of allergy shots. Some evidence suggests that the IgG4 antibodies have a 'blocking' effect. This can occur at two levels. In the first, binding to allergen at mucosal surfaces before it has a chance to interact with FcεRI-bound IgE inhibits immediate hypersensitivity. IgG4 may also blunt T_H2 responses by inhibiting 'facilitated antigen presentation' by tissue antigen-presenting cells. These cells can carry allergen-specific IgE bound to its receptors, FcεRI or CD23, and can thus take up allergen efficiently.

It is also possible that the induction of IgG4 responses reflects an important shift in allergen-specific T-cell responses in the nasal mucosa. In allergic subjects, antigen-specific T cells have a T_H2 phenotype (producing IL-4, IL-5, and IL-13), which supports local IgE production, eosinophil recruitment, and T_H2 expansion. A consistent effect of allergen immunotherapy is the expansion of regulatory T cells, along with their signature cytokines, IL-10 and TGF-β. IL-10 is detectable in the respiratory mucosa of subjects treated with allergy shots and has several anti-allergic effects, including suppression of IgE production by B cells (and induction of IgG4), inhibition of FcεRI-triggered mast-cell activation, and suppression of T_H2 cytokine production. The decrease in IgE production not only reduces the major trigger of allergen-driven mast-cell activation but also has significant effects on mast-cell homeostasis. IgE antibodies promote the growth and survival of tissue mast cells. Reduction of IgE levels can lead to decreased mast-cell numbers in allergen-exposed tissues and a consequent reduction in the pool of allergic mediators.

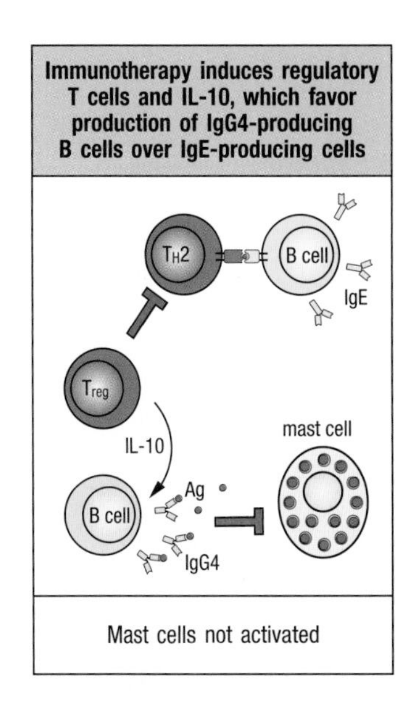
Immunotherapy induces regulatory T cells and IL-10, which favor production of IgG4-producing B cells over IgE-producing cells
TH2
B cell
IgE
Treg
IL-10
mast cell
Ag
B cell
IgG4
Mast cells not activated

Fig. 3.7 Mechanisms of allergen immunotherapy. Immunotherapy induces regulatory T cells. Regulatory T cells (T_{reg}) produce IL-10, a cytokine that inhibits T_H2 activation, IgE production by B cells, and the release of mediators by mast cells. IgG4 production is enhanced, however. IgG4 antibodies may have 'blocking' functions, inhibiting the interaction of their cognate antigens with IgE bound on mast cells and antigen-presenting cells, thereby blocking the release of mediators of immediate hypersensitivity and the process of facilitated antigen presentation.

Questions.

1 *Why does Charlie have allergic symptoms all day during the springtime, even when he spends the evening inside?*

2 *Patients such as Charlie with nasal allergies frequently run into difficulties with recurrent sinus infections. How does allergic rhinitis increase the risk of developing sinusitis?*

3 *What might be the risks of administering immunotherapy?*

CASE 4 Allergic Conjunctivitis

Allergic reactions in the mucous membranes of the eye.

The eyes are constantly exposed to aeroallergens. Tears bathing the conjunctivae, the mucous membranes lining the eyes and the inner surfaces of the eyelids, form a film laden with environmental proteins. The conjunctival mucosa is therefore a major site both of immunological sensitization to environmental proteins and of allergic reactions. Hypersensitivity responses in the conjunctivae can arise by several mechanisms. According to the classic scheme developed by Gell and Coombs in the early 1960s, allergic reactions can be mediated by four major pathways (Fig. 4.1). In the type I reaction, IgE-mediated mast-cell activation via the high-affinity IgE receptor FcεRI leads to the release of preformed mast-cell mediators, including proteases, proteoglycans, and histamine, as well as the rapid synthesis of arachidonic acid metabolites including prostaglandins (PGD_2 and PGE_2), and leukotrienes (including LTC_4, LTD_4, and LTE_4) (Fig. 4.2). These mediators of immediate hypersensitivity act on vascular endothelium and nerve endings in the conjunctival mucosa to induce immediate vascular leak and edema (swelling and congestion) as well as pruritis (itching) and tearing. In contrast, chronic forms of conjunctival inflammation are mediated by a combination of type I and type IV (cell-mediated) hypersensitivity mechanisms. In many circumstances the conjunctival allergic response to allergens involves more than one hypersensitivity mechanism. For example, cytokines (interleukin (IL)-4, IL-5, IL-13, and tumor necrosis factor (TNF)-α) produced by IgE-activated mast cells a few hours after FcεRI signaling drive the local expression of vascular adhesion molecules which, along with chemokines (CCL11 (eotaxin) and CCL24), attract eosinophils and T_H2 cells into tissues, setting up the late-phase reaction that typically follows 8–12 hours after initial allergen encounter (Fig. 4.3). Over time and with recurrent allergen exposures, T_H2 cells recruited to the conjunctival mucosa drive a chronic allergic inflammatory response and support ongoing local IgE production by B cells residing in the mucosa. In this way, a type I reaction can ultimately promote the induction of type IV responses.

This case was prepared by Hans Oettgen, MD, PhD, and Raif Geha, MD, in collaboration with James Friedlander, MD.

Topics bearing on this case:

- IgE antibodies
- Hypersensitivity reactions
- Mast-cell activation

	Type I	Type II		Type III	Type IV		
Immune reactant	IgE	IgG		IgG	T_H1 cells	T_H2 cells	CTL
Antigen	Soluble antigen	Cell- or matrix-associated antigen	Cell-surface receptor	Soluble antigen	Soluble antigen	Soluble antigen	Cell-associated antigen
Effector mechanism	Mast-cell activation	Complement, FcR^+ cells (phagocytes, NK cells)	Antibody alters signaling	Complement, phagocytes	Macrophage activation	IgE production, eosinophil activation, mastocytosis	Cytotoxicity
	Ag	platelets + complement		immune complex; blood vessel; + complement	IFN-γ; T_H1; chemokines, cytokines, cytotoxins	IL-4; IL-5; T_H2; eotaxin; cytotoxins, inflammatory mediators	CTL
Example of hypersensitivity reaction	Allergic rhinitis, allergic asthma, atopic eczema, systemic anaphylaxis, some drug allergies	Some drug allergies (e.g., penicillin)	Chronic urticaria (antibody against FcεRI alpha chain)	Serum sickness, Arthus reaction	Allergic contact dermatitis, tuberculin reaction	Chronic asthma, chronic allergic rhinitis	Graft rejection, allergic contact dermatitis to poison ivy

Fig. 4.1 Mechanisms of immunologically mediated tissue damage. Type I hypersensitivity responses are mediated by IgE, which induces mast-cell activation. Type II responses are IgG-mediated reactions to cell-surface antigens, whereas type III responses, which are also IgG-mediated, target soluble antigens, resulting in immune complex formation. Type IV hypersensitivity reactions are T-cell mediated and can be subdivided into three groups. In the first group, tissue damage is caused by the activation of macrophages by T_H1 cells, resulting in an inflammatory response. In the second group, allergic tissue inflammation is induced by the activation by T_H2 cells. In the third group, damage is caused directly by cytotoxic T cells (CTL).

The case of Quentin Compson: an 8-year-old boy with itchy eyes.

Quentin Compson was referred to the Children's Hospital Allergy Program by his primary care physician in April at the age of 8 years. His mother told the allergist that Quentin had been rubbing his eyes frequently over the preceding few weeks. He complained of intense itching and had resorted to carrying tissues to wipe away his constant tears. His symptoms were worse while outdoors and most pronounced in the morning. Quentin's mother had tried giving him oral diphenhydramine, a histamine blocker, which only partly relieved his symptoms. He had become quite drowsy on this medication, and his mother was concerned about his going to school in this condition.

From Quentin's medical history, it became evident that he had experienced similar symptoms around the same time the previous year, but that his eyes had been fine in the interim. His mother also mentioned that he had developed itchy, watery eyes on several occasions while visiting his maternal aunt, who has two cats. A review of his symptoms revealed that Quentin had had nasal congestion, clear nasal discharge, frequent sneezing, and an itchy throat over the past 2 weeks. There was no fever or cough, and he had not complained of photophobia (extreme sensitivity to light) or changes in his vision. A family history revealed that Quentin's 3-year-old sister had eczema and a history of wheezing with viral infections. His father had asthma and allergic rhinitis.

8-year-old boy with
redness, itching, and
tearing of the eyes.

Physical examination revealed a boy who appeared well but was rubbing his eyes, which had a clear, watery discharge. The sclerae of both eyes were injected (red) all

Class of product	Examples	Biological effects
Enzyme	Tryptase, chymase, cathepsin G, carboxypeptidase	Remodel connective tissue matrix
Toxic mediator	Histamine, heparin	Toxic to parasites Increase vascular permeability Cause smooth muscle contraction Anticoagulation
Cytokine	IL-4, IL-13	Stimulate and amplify T_H2-cell response
	IL-3, IL-5, GM-CSF	Promote eosinophil production and activation
	TNF-α (some stored preformed in granules)	Promotes inflammation, stimulates cytokine production by many cell types, activates endothelium
Chemokine	CCL3	Attracts monocytes, macrophages, and neutrophils
Lipid mediator	Prostaglandins D_2, E_2 Leukotrienes C_4, D_4, E_4	Smooth muscle contraction Chemotaxis of eosinophils, basophils, and T_H2 cells Increase vascular permeability Stimulate mucus secretion Bronchoconstriction
	Platelet-activating factor	Attracts leukocytes Amplifies production of lipid mediators Activates neutrophils, eosinophils, and platelets

Fig. 4.2 Molecules released by mast cells on activation. Mast cells produce a wide variety of biologically active proteins and other chemical mediators. The enzymes and toxic mediators listed in the first two rows are released from the preformed granules. The cytokines, chemokines, and lipid mediators are synthesized after activation.

over. The pupils were equal, round, and reactive to light, and extraocular movements were intact. Fundoscopy showed sharp optic disks, and no foreign body was seen. The tympanic membranes were translucent and the nasal mucosa was boggy and pale with enlarged nasal turbinates. Breath sounds were clear bilaterally. There was no rash.

After his physical exam, Quentin underwent skin-prick testing to a panel of environmental allergens (see Case 3). He had positive results for birch, oak, elm, and maple tree pollens, as well as for orchard and timothy grass pollens. He also tested positive

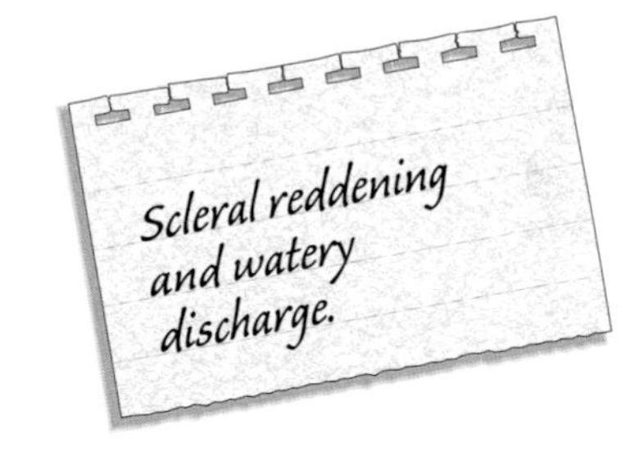

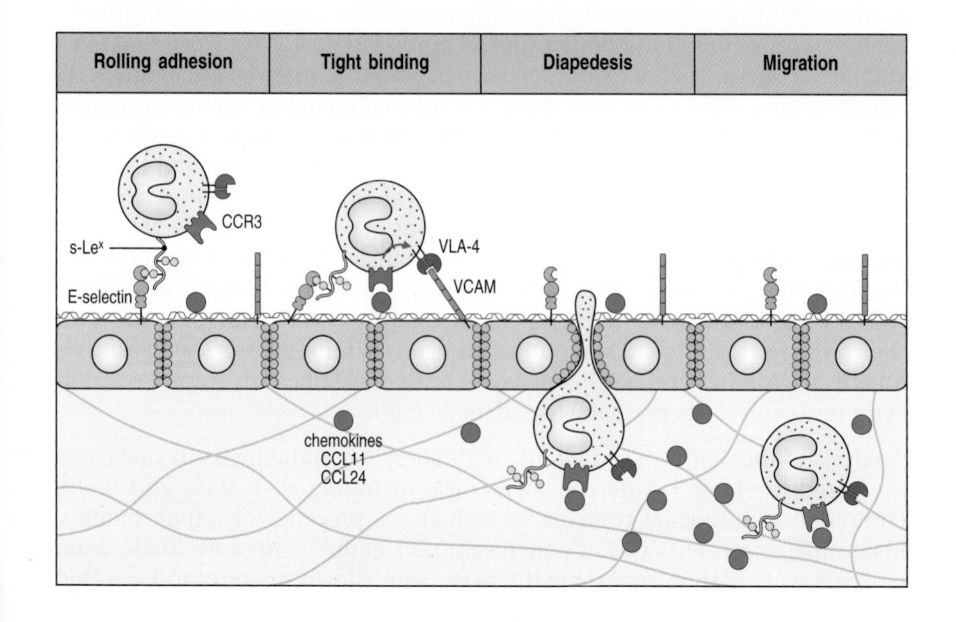

Fig. 4.3 Eosinophil recruitment in an allergic reaction. Hours after allergen inhalation, eosinophils are recruited from the bloodstream into the nasal mucosa. Cytokines, including TNF-α, produced upon the IgE-mediated activation of mast cells, induce the expression of adhesion molecules called selectins in the vascular endothelium. Brief, reversible interactions between E- and P-selectins and their corresponding receptors on eosinophils lead to rolling of the cells along the endothelium. Subsequent tight interactions between the eosinophil integrin VLA-4 and its endothelial ligand VCAM lead to the arrest of rolling. Eosinophil-directed chemokines, including CCL11 (eotaxin) and CCL24, then promote transmigration (diapedesis) of the eosinophils across the vascular endothelium and into the mucosa. S-Le^x: Sialyl Lewis^x.

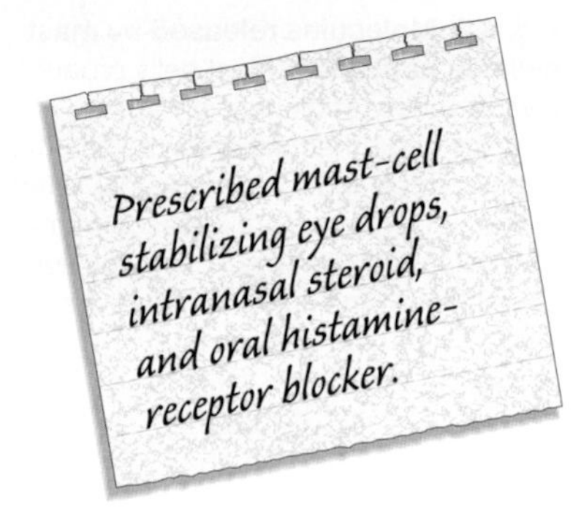
Prescribed mast-cell stabilizing eye drops, intranasal steroid, and oral histamine-receptor blocker.

to dust mites and cat dander. A complete blood count with differential revealed a normal white blood cell count with a mildly elevated absolute eosinophil count.

Quentin was treated with three medications. For his eye symptoms, he was prescribed ketotifen (a mast-cell stabilizer) ophthalmic solution, one drop in each eye twice daily. For his nasal congestion and rhinorrhea, he was given fluticasone (a glucocorticoid) nasal spray, with instructions to use two sprays in each nostril once daily. He was also advised to use oral cetirizine (a histamine-receptor blocker). His family was instructed on how to avoid exposure to allergens. Quentin was advised to avoid prolonged outdoor exposure during the peak pollen periods.

Allergic conjunctivitis.

Itchy red eyes are a common and bothersome affliction among people sensitive to environmental allergens. Allergic eye diseases can be classified into five types: seasonal allergic conjunctivitis, which was the type diagnosed in Quentin, perennial allergic conjunctivitis, vernal keratoconjunctivitis, atopic keratoconjunctivitis, and giant papillary conjunctivitis.

In seasonal and perennial allergic conjunctivitis the major symptom is ocular itching, which can range from mild to severe. Additional symptoms include profuse tears, ocular redness, burning, swelling of the eyelids, and a sensation of fullness around the eyelid and periorbital areas. Patients may complain of photophobia and/or blurry vision, though these symptoms are less common. Symptoms of allergic rhinitis, including nasal congestion and clear nasal discharge (see Case 3), often accompany those of allergic conjunctivitis. Physical examination reveals conjunctival reddening and edema. The edema is the result of increased permeability of the postcapillary venules. Often, patients will develop 'allergic shiners,' dark areas under the eyes resulting from impaired venous return from the skin and subcutaneous tissues. The diagnosis is made clinically on the basis of the patient's history and physical examination. Skin-prick and intradermal testing to environmental allergens can subsequently be used to identify the responsible allergens. Testing for allergen-specific IgE in the serum can be helpful in diagnosis in some cases.

As in Quentin's case, treatment of seasonal allergic conjunctivitis with oral antihistamines is often unsatisfactory, failing to provide adequate symptomatic relief. Current recommendations suggest the use of dual-action ophthalmic drops containing both a topical antihistamine as well as a mast-cell stabilizing agent, such as ketotifen, which prevents mast-cell activation. As would be predicted given that mast cell-derived mediators drive eosinophil trafficking to the conjunctival mucosa, the dual-action topical medications have additionally been shown to inhibit eosinophil infiltration and the late-phase inflammatory response. These topical ocular medications have a secondary beneficial effect on nasal symptoms in patients with allergic rhinoconjunctivitis, perhaps due in part to their transfer to the nose via the nasolacrimal duct, a direct connection between eyes and nose that normally serves to clear tears. In many patients these medicines provide adequate relief. In those who have persistent symptoms, allergen immunotherapy (see Case 3) is an effective option and has been shown both to significantly improve ocular symptoms and to decrease the use of medication.

Analysis of the tear film in patients with allergic conjunctivitis has revealed increased levels of T_H2-derived cytokines, including IL-4, IL-5, and IL-13, compared with normal controls, as well as the presence of IgE antibodies, histamine, tryptase, CCL11, and eosinophil cationic protein. Tears from patients with allergic conjunctivitis have been shown to contain decreased amounts of the anti-inflammatory cytokine IL-10, compared with healthy individuals, and decreased levels of interferon-gamma (IFN-γ), which is produced by T_H1 cells.

Although most patients with Quentin's symptoms have either seasonal or perennial allergic conjunctivitis, it is important to consider other forms of ocular allergy, because some can permanently affect vision. The differential diagnosis of a patient with allergic conjunctivitis includes the rarer diseases—vernal keratoconjunctivitis, atopic keratoconjunctivitis, and giant papillary conjunctivitis.

Vernal keratoconjunctivitis is a chronic sight-threatening inflammatory disorder that is more common in children living in warm climates. The onset of symptoms occurs in the spring months some time in the first decade of life. In many patients, the disease disappears during puberty. Predominant symptoms include severe photophobia and intense ocular itching. As patients with vernal keratoconjunctivitis experience recurrent conjunctivitis in the spring, it is often confused with seasonal allergic conjunctivitis. There are, however, several key distinguishing features. Most helpful diagnostically is the growth of giant papillae on the tarsal (lining the eye lids) conjunctivae, which take on a cobblestone appearance (Fig. 4.4). These papillae are infiltrated with lymphocytes and eosinophils (Fig. 4.5). Physical examination may also reveal a thick, ropy, mucous discharge on the tarsal conjunctivae. 'Horner–Trantas' dots, which are collections of eosinophils and epithelial cells, can be seen around the limbus (the edge of the cornea where it borders the sclera). It is thought that the T_H2-type cytokines IL-4 and IL-13 stimulate migration, proliferation, and collagen production by fibroblasts, perhaps via the 'alternative' pathway of macrophage activation (see Case 14), as well as the formation of giant papillae. If untreated, patients with vernal keratoconjunctivitis can develop corneal erosions, which may progress to ulcers. The disease requires immediate and aggressive treatment, usually involving close collaboration between allergy and ophthalmology specialists. Treatment includes topical corticosteroids, antihistamines, and mast-cell stabilizing compounds. Topical corticosteroids, although effective, can have significant side effects, including an increase in intraocular pressure, cataracts, and an increased risk of infection.

Eyelid papillae in vernal keratoconjunctivitis

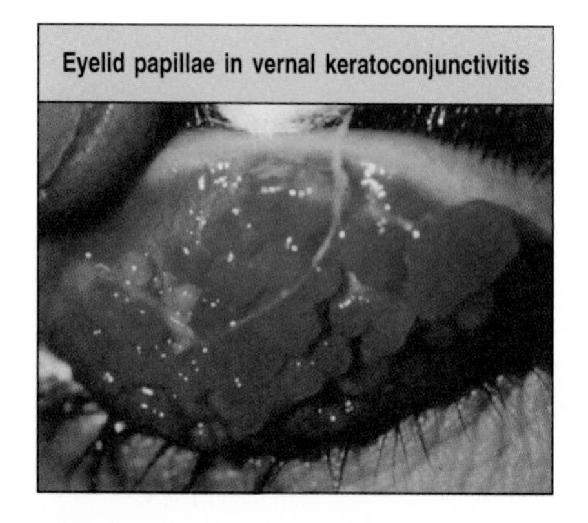

Fig. 4.4 Eyelid papillae in vernal keratoconjunctivitis. In vernal keratoconjunctivitis (VKC), the eyelid papillae have a cobblestone appearance. T_H2 responses drive collagen production by fibroblasts and influence the formation of papillae. Patients with VKC also can have a thick, ropy, mucous discharge from the eye.

Atopic keratoconjunctivitis is a severe chronic inflammatory ocular disorder affecting both the conjunctivae and the outer eyelids of patients with atopic dermatitis (see Case 5). It occurs mostly in adults and is more common in men. A family history of atopic disease is usually present. Patients experience intense ocular itching, burning, and tearing. The eyelids of patients with atopic keratoconjunctivitis are scaly, thickened, and lichenified, consistent with atopic dermatitis. Keratoconus, or non-inflammatory thinning of the cornea, is present in some patients. Symptoms can be present all year round, with seasonal exacerbations common. Severe complications from the disease include keratopathy, corneal ulceration, cataracts, and retinal detachment. In a large majority of patients with this type of keratoconjunctivitis, the conjunctivae are colonized by the bacterium *Staphylococcus aureus*, which may occur because of defects in innate immunity, including the impaired production of antimicrobial peptides as occurs in atopic dermatitis. Treatment of atopic keratoconjunctivitis includes antihistamine and mast-cell stabilizing medications as well as topical steroids and other topical immunosuppressive treatments.

Shave conjunctival papillary biopsy

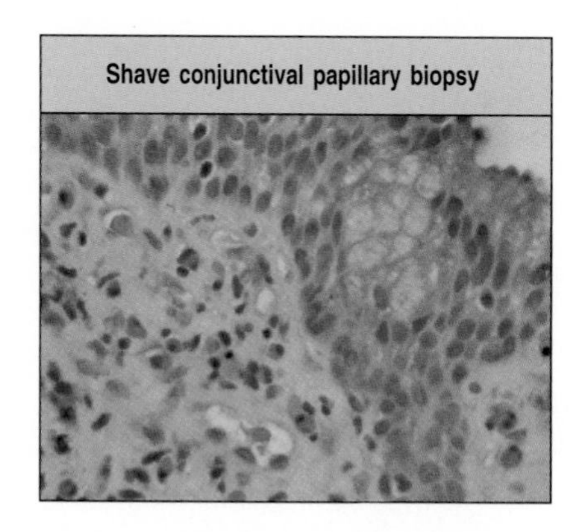

Fig. 4.5 A shave conjunctival papillary biopsy from a patient with vernal keratoconjunctivitis. A shave biopsy was used to remove a very superficial layer of conjunctival tissue from a patient with vernal keratoconjunctivitis. This type of biopsy does not require suturing or other repair. Note the abundance of lymphocytes (blue staining) and eosinophils (orange-pink cytoplasmic staining).

Giant papillary conjunctivitis is a chronic inflammatory condition of the eyes commonly seen in patients who use extended-wear contact lenses or other types of ocular prostheses. Immunologically, this type of conjunctivitis is thought to involve inflammation of external ocular structures, induced by a cellular delayed-type hypersensitivity response (see Fig. 4.1), not by IgE-mediated mechanisms, and it is not sight-threatening. Contact lens polymers, thimerosal, and protein deposits on contact lens surfaces have all been implicated as potential antigens driving the immune response in giant papillary conjunctivitis, but the etiology is still unclear. Symptoms can mimic those of seasonal or perennial allergic conjunctivitis, including ocular itching and irritation, which progress with time and can be worse upon insertion of the contact lens. Patients develop striking giant papillae on the conjunctivae (see Fig. 4.4), which creates a cobblestone appearance. Although the condition can be treated with appropriate contact lens hygiene, artificial tears, topical antihistamines, mast-cell stabilizers, and even topical corticosteroids, temporary removal of the contact lenses is usually necessary. The prognosis is good, with most patients able to return to contact lens use after an appropriate treatment.

The IgE-mediated processes driving seasonal and perennial allergic conjunctivitis do not account for the chronic and severe symptoms seen in the more severe forms of ocular allergy. Vernal keratoconjunctivitis, for example, is thought to arise from a combination of type I and type IV hypersensitivity reactions. Patients with this disease have increased numbers of T_H2 cells, mast cells and eosinophils, and increased levels of T_H2-derived cytokines, but allergen-specific IgE antibodies are detected in only 50% of patients, which is consistent with IgE-independent pathways of T_H2-driven allergic inflammation.

Questions.

1 *Why might oral antihistamines fail to provide adequate relief to a patient suffering from seasonal allergic rhinitis?*

2 *Which physical exam findings can help distinguish between seasonal allergic conjunctivitis and vernal keratoconjunctivitis?*

3 *Describe the pathway from contact with an allergen to the development of symptoms in a patient with seasonal allergic conjunctivitis.*

4 *Quentin's allergies were identified by skin testing. Under what circumstances would this type of testing be of limited value, and what alternative approach could be used?*

CASE 5 Atopic Dermatitis

Skin as a target organ for allergic reactions.

As we saw in Cases 1 and 2, allergic or hypersensitivity reactions to otherwise innocuous antigens occur in certain individuals. The site of such reactions and the symptoms they produce vary depending on the type of allergen and the route by which it enters the body. Here we consider reactions to allergens entering the skin.

The main function of skin is to provide a physical barrier to the entry of foreign materials—including irritants, allergens, and pathogens—and to regulate the loss of water. The skin is composed of three layers—the epidermis (the outermost layer), the dermis, and the hypodermis—and the barrier function is performed by the epidermis. The epidermis consists mainly of keratin-synthesizing stratified epithelial cells—keratinocytes—in various stages of differentiation, interspersed with antigen-presenting Langerhans cells, a type of immature dendritic cell. The outermost cornified layer of the epidermis—the stratum corneum—consists of dead keratinocytes (corneocytes) filled with fibrous proteins such as keratin and involucrin. The dermis is a vascularized layer consisting of fibroblasts and dense connective tissue with collagen and elastic fibers, populated by dendritic cells, mast cells, macrophages, and few lymphocytes. The hypodermis is a layer of fat cells and loose connective tissue in contact with the underlying muscle.

After an antigen has been introduced into the skin, it is captured by Langerhans cells and other dendritic cells; these then migrate to local draining lymph nodes, where they present the antigen to recirculating naive T cells. Activated antigen-specific T cells proliferate and differentiate into memory or effector cells that express skin-homing receptors such as cutaneous lymphocyte antigen (CLA), and the chemokine receptors CCR4 and CCR10. CLA, which is an inducible carbohydrate modification of P-selectin glycoprotein ligand-1, is almost absent on naive T cells but is expressed by 30% of circulating memory T cells and by approximately 90% of infiltrating T cells in inflamed skin. CLA^+ T cells coexpress CCR4, whose ligands are the chemokines CCL17 (TARC) and CCL22 (MDC). Chemokines are small polypeptides that are synthesized by many cells, including keratinocytes (skin epidermal cells), fibroblasts, effector T cells, eosinophils, and macrophages (Fig. 5.1). They act through receptors that are members of the G-protein-coupled seven-span receptor family.

This case was prepared by Raif Geha, MD.

Topics bearing on this case:

- Cytokines and chemokines
- Inflammatory responses
- Bacterial superantigens
- T_H2 cell responses
- Migration and homing of lymphocytes
- IgE and allergic reactions

Fig. 5.1 Properties of selected chemokines. Chemokines fall into three related but distinct structural subclasses: CC chemokines have two adjacent cysteines at a particular point in the amino acid sequence; CXC chemokines have the same two cysteine residues separated by another amino acid; and C chemokines (not shown) have only one cysteine residue at this site.

Chemokine class	Chemokine	Produced by	Receptors	Major effects
CXC	CXCL8 (IL-8)	Monocytes Macrophages Fibroblasts Keratinocytes Endothelial cells	CXCR1 CXCR2	Mobilizes, activates, and degranulates neutrophils Angiogenesis
CC	CCL3 (MIP-1α)	Monocytes T cells Mast cells Fibroblasts	CCR1, 3, 5	Competes with HIV-1 Anti-viral defense Promotes T_H1 immunity
	CCL4 (MIP-1β)	Monocytes Macrophages Neutrophils Endothelium	CCR1, 3, 5	Competes with HIV-1
	CCL2 (MCP-1)	Monocytes Macrophages Fibroblasts Keratinocytes	CCR2B	Activates macrophages Basophil histamine release Promotes T_H2 immunity
	CCL5 (RANTES)	T cells Endothelium Platelets	CCR1, 3, 5	Degranulates basophils Activates T cells Chronic inflammation
	CCL11 (Eotaxin)	Endothelium Monocytes Epithelium T cells	CCR3	Role in allergy

Antigen-specific effector CD4 T cells leave the draining lymph nodes and are returned to the bloodstream, from which they reenter the skin via their receptors for skin-specific adhesion molecules and chemokines. In the skin, they can be activated locally by specific antigen to proliferate further and secrete effector cytokines.

When an allergen enters skin whose normal barrier function is disrupted in some way, it elicits an immune response dominated by allergen-specific T_H2 cells secreting their characteristic cytokines (for example, interleukin-4 (IL- 4)), and an IgE antibody response. This T_H2-dominated immune response results in a chronic skin inflammation and localized tissue destruction called atopic dermatitis or atopic eczema—to distinguish it from cases of dermatitis that do not have an allergic basis.

A chronic inflammatory reaction is sustained by the lymphocytes, eosinophils, and other inflammatory cells that are attracted out of the blood vessels at the site of inflammation. The first step in the process of lymphocyte homing to skin is the reversible binding (rolling) of lymphocytes to the vascular endothelium through interactions between CLA on the lymphocyte with E-selectin on the endothelial cells. The rolling cells are brought into contact with chemokines retained on heparan sulfate proteoglycans on the endothelial cell surface. Chemokine signaling activates the lymphocyte integrin lymphocyte function-associated antigen-1 (LFA-1), leading to firm adherence to intercellular cell adhesion molecule-1 (ICAM-1) on the endothelium followed by extravasation, or departure from the blood vessel, into the skin. Eosinophils migrate into tissues in a similar way, via an interaction between the integrin very late antigen-4 (VLA-4) on the eosinophil and vascular cell adhesion molecule-1 (VCAM-1) on the vascular endothelium. Once the lymphocytes and other cells have crossed the endothelium into the dermis, their migration to the focus of inflammation is directed by a gradient of chemokine molecules bound to the extracellular matrix.

If effector T cells are activated by their specific antigen once they have reentered the skin, they produce chemokines such as CCL5 and cytokines such as TNF-α (which activates endothelial cells to express E-selectin, VCAM-1, and ICAM-1). The chemokines produced by the effector T cells—and, under their influence, by keratinocytes—act on other T cells to upregulate their adhesion molecules, thus increasing recruitment of T cells into the affected tissue. At the same time, monocytes and polymorphonuclear leukocytes are recruited to these sites by adhesion to E-selectin. The TNF-α and interferon (IFN)-γ released by the activated T cells also act synergistically to change the shape of endothelial cells, resulting in increased blood flow, increased vascular permeability, and increased immigration of leukocytes, fluid, and protein into the site of inflammation. Thus, a few allergen-specific T cells encountering antigen in a tissue can initiate and amplify a potent local inflammatory response that recruits both antigen-specific and accessory cells to the site.

The case of Tom Joad: the itch that rashes.

Tom was admitted to the hospital when he was 2 years old because of his worsening eczema. In the week before admission he had developed many open skin lesions (erosions), increased itching (pruritis), redness (erythema), and swelling (edema) of the skin. The lesions oozed a clear fluid, which formed crusts around them.

2-year-old boy with
severe eczema and
family history of
allergy.

Tom had suffered from eczema since the age of 2 months, when he developed a scaly red rash over his cheeks and over his knees and elbows (Fig. 5.2). He was breast-fed until 3 months old, when he was given a cow's milk-based formula. After 24 hours on the formula he started to vomit and to scratch his skin. A casein hydrolyzate formula was substituted for milk and he tolerated this well, but as new foods were added to his diet the eczema worsened. At 9 months old he developed a wheeze and was treated with bronchodilators. At 2 years old he had hives after eating peanut butter.

Tom's mother suffers from hay fever and his father has atopic dermatitis. The family lives in an old house with 20-year-old carpeting. Tom slept on a 10-year-old mattress surrounded by lots of stuffed animals. There were no pets, and his parents did not smoke.

During physical examination Tom was evidently uncomfortable and scratched continuously at his skin. His temperature was 37.9°C, pulse 96 beats min^{-1}, respiratory rate 24 min^{-1}, blood pressure 98/58 mmHg, weight 12 kg (10th centile), and height 90 cm (25th centile). His skin was very red, with large scales, and with scratched and infected lesions on his face, trunk, and extremities. Pustules were present on his arms and legs, and there were thick scales in his scalp. Thickened plaques of skin (lichenification) with a deep criss-cross pattern were seen around the creases on the insides of his elbows and knees, and on the backs of his hands and feet.

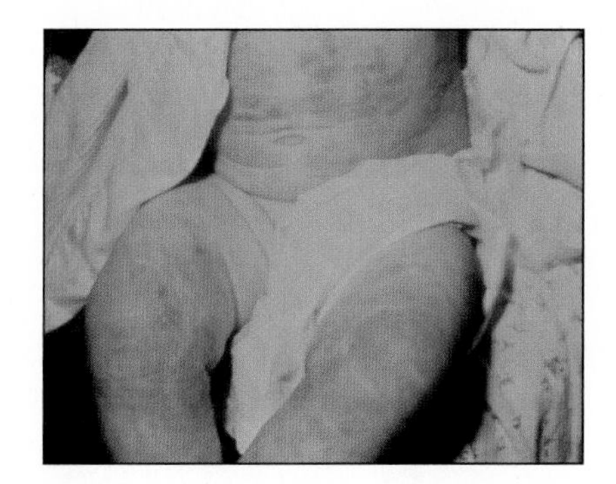

Fig. 5.2 An infant with severe eczema due to atopic dermatitis. Note the reddened and broken skin, especially in places, such as over the knees, where it is subjected to continual stretching and stress. Photograph courtesy of S. Gellis.

A skin culture was positive for *Staphylococcus aureus* and *Streptococcus pyogenes* Group A. Tom was treated with intravenous oxacillin (an antibiotic), antihistamines, topical steroids, and skin emollients such as coal tar. The infection resolved and his skin healed. Laboratory studies during hospitalization showed a white blood cell count of 9600 μl^{-1}, with 41% polymorphonuclear leukocytes, 26% lymphocytes, and 25% eosinophils (normal 0–5%), 13.1 g dl^{-1} hemoglobin, and a hematocrit of 37.2%. The absolute eosinophil count was elevated at 2400 μl^{-1} (normal 0–500 μl^{-1}). Serum IgE was also much elevated at 32,400 IU ml^{-1} (normal 0–200 IU ml^{-1}).

After discharge from hospital, Tom attended the allergy clinic, where he was tested for sensitivity to a range of allergens. He showed a positive type I allergic skin response (see Fig. 2.5) to numerous inhaled allergens including dust mites, mold spores, animal dander, and a variety of pollens. Tom also tested positive in a type I skin response to milk, cod, wheat, egg white, peanut, and tree nuts (cashew, almond, pecan, walnut, Brazil nut, and hazel nut) but had no reaction to rice or soybean. To determine whether any of these foods could be causing his atopic dermatitis, double-blind placebo-controlled food challenges were performed. He developed hives and wheezing after eating 1 g of egg white, and hives and eczema after drinking 2 g of powdered milk, but had no reaction to wheat or to cod. Four hours after the milk challenge, his serum tryptase level was raised, indicating IgE-induced mast-cell degranulation.

His parents were advised to cover his mattress and pillows with a plastic covering and to remove the carpet and stuffed animals from his bedroom to decrease exposure to dust and mite allergens. Tom had a history of allergic reaction to peanuts, so continued avoidance of peanuts and tree nuts was recommended; he was therefore placed on a diet that excluded milk, eggs, peanuts, and tree nuts.

Reduce exposure to allergens.

Tom's eczema improved significantly in response to these environmental and dietary control measures, together with the use of emollients and low-potency topical steroids on his skin. The family reduced the risk of skin irritation by avoiding all perfumed soaps and lotions, double-rinsing Tom's laundry, and dressing him in cotton clothes. He did well on this regimen until he was 12 years old, when he awoke one day with itchy vesicles on his lower left leg and ankle. The lesions progressed to become painful punched-out erosions, and were diagnosed as herpes simplex infection. He was given the antiviral drug acyclovir and the lesions resolved. Tom is now 15 years old.

Atopic dermatitis.

Atopic dermatitis, or atopic eczema, is a common pruritic (itching) inflammatory skin disease often associated with a family and/or personal history of allergy. Its prevalence is currently about 17% in children in the United States, and it is on the rise all over the world but particularly in Western and industrialized societies. Atopic dermatitis almost uniformly starts in infancy and although it tends to resolve, or improve remarkably, by the age of 5 years, it can persist into adult life in about 15% of cases. Many children with atopic dermatitis develop other indications of atopy—a predisposition to develop allergies—such as food allergies, asthma, and allergic rhinitis (hay fever).

The hallmark of atopic dermatitis is skin barrier dysfunction, which results in dry itchy skin. This leads to scratching, which inflicts mechanical injury and allows access to environmental antigens, resulting in sensitization and allergic skin inflammation (Fig. 5.3). The barrier to water permeation even in normal skin is not absolute, and the movement of water through the stratum corneum into the atmosphere is known as transepidermal water loss (TEWL). Almost all patients with atopic dermatitis have increased TEWL, an indicator of disrupted barrier function.

The terminal differentiation of keratinocytes from granular cells to corneocytes is a critical step in the maintenance of skin barrier function. The formation of the stratum corneum involves the sequential expression of several major structural proteins. Many of the proteins involved in skin cornification are encoded in a locus containing about 70 genes on chromosome 1q21, termed the epidermal differentiation complex (EDC). EDC genes are expressed during the late stages of terminal keratinocyte differentiation and encode proteins such as loricrin, involucrin, and the S100 calcium-binding

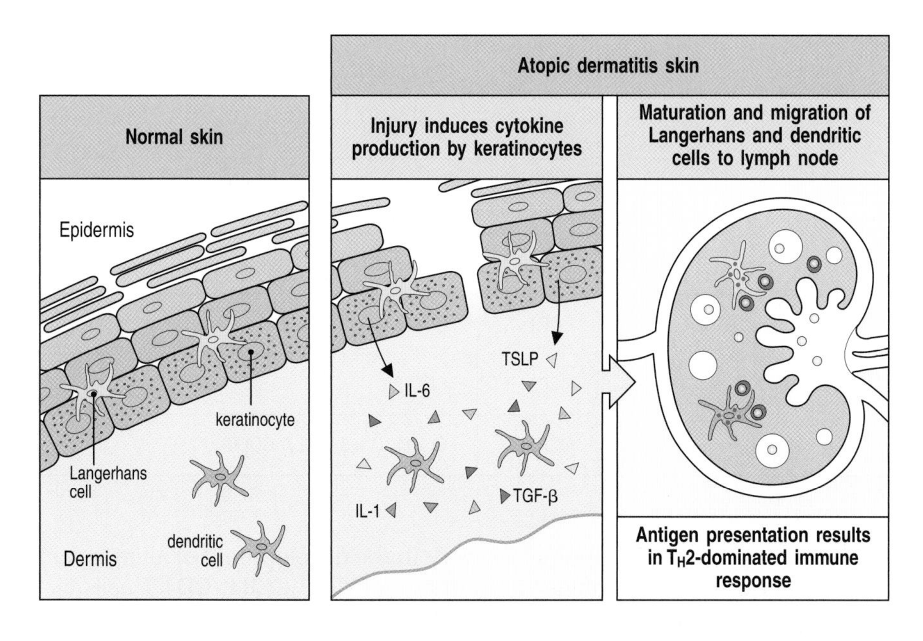

Fig. 5.3 Skin barrier dysfunction in atopic dermatitis. An intact skin barrier (left panel) prevents allergens from entering normal skin. Damage to this barrier (right panels) allows allergens to penetrate into the subepidermal layer and interact with antigen-presenting cells, and induces keratinocytes to produce cytokines that include TSLP, IL-1, IL-6, and TGF-β. This leads to maturation and migration of the antigen-presenting cells to the draining lymph nodes, where they present antigens to naive T cells, resulting in a T_H2-dominated immune response.

protein filaggrin. During corneocyte differentiation, the giant inactive precursor polypeptide profilaggrin is dephosphorylated and cleaved by serine proteases into multiple filaggrin polypeptides, which are further degraded to hydrophilic amino acids that have a critical role in the hydration of the skin.

The filaggrin gene is mutated in about 20% of patients with atopic dermatitis, and this association suggests that the skin-barrier defect precedes the development of the disease. Evidence for a role of defective skin barrier in atopic dermatitis is provided by the observation that application of the protein ovalbumin to mouse skin that has been disrupted by tape-stripping results in an allergic inflammation of the skin that has features of atopic dermatitis. Because *filaggrin* mutations are present in only a fraction of patients with atopic dermatitis, however, genetic variants of other genes involved in the skin's barrier function are also likely to be important in the pathogenesis of the disease. Indeed, mutations in two other genes involved in skin barrier function have been associated with atopic dermatitis, namely *SCCE* and *SPINK5*.

It is not clear why the introduction of antigen through a disrupted skin barrier results in a T_H2-dominated allergic response. However, it is known that mechanical injury causes keratinocytes to release the epithelial cytokine thymic stromal lymphopoietin (TSLP). TSLP acts on dendritic cells to promote their ability to skew naive T cells toward differentiation to T_H2 cells. TSLP levels are increased in skin lesions of atopic dermatitis, and expression of a TSLP transgene in keratinocytes in mice results in allergic skin inflammation. TSLP also acts on skin-infiltrating effector T helper cells to promote T_H2 cytokine production.

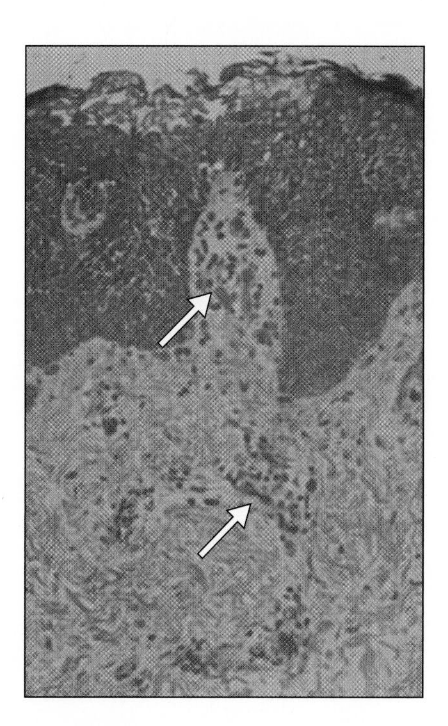

Fig. 5.4 A section through an acute skin lesion from a patient with atopic dermatitis. The section is stained with hematoxylin and eosin. The dermis has been infiltrated by mononuclear cells (arrowed), which are mostly T cells.

The skin lesions in atopic dermatitis contain a mononuclear cell infiltrate that is predominantly located in the dermis (Fig. 5.4) and is composed of activated memory CD4 T cells and macrophages. The T cells involved in the skin lesions of acute atopic dermatitis are principally skin-homing CLA^+ $CCR4^+$ T_H2 cells. The T_H2-cell cytokines IL-4, IL-13, and IL-5 in particular have key roles in the condition. IL-4 and IL-13 cause skin-resident keratinocytes and dermal fibroblasts to secrete the chemokines CCL5 (RANTES), CCL11, and CCL22, ligands for CCR3 and CCR4 expressed on skin-homing T cells; CCL11 (eotaxin), which attracts eosinophils; and CCL13 (MCP-4), which attracts macrophages. There is therefore an influx of T cells, eosinophils, and macrophages into the skin. The chemokines and receptors known to be involved in atopic dermatitis are listed in Fig. 5.5. The selective homing of memory or effector CD4 T cells to

Fig. 5.5 Chemokines that act on T cells and eosinophils.

Target cell		Receptor	Chemokine
T cells	T_H1	CXCR3	CXCL10 (IP-10), CXCL9 (MIG)
	T_H2	CCR3 CCR4	CCL11, CCL5, CCL7, CCL13 (MCP-3, MCP-4), CCL5, CCL3, CCL2, CCL17 (TARC), CCL22 (MDC)
	Both	CCR10	CCL27 (CTACK), CCL28
Eosinophils		CCR1	CCL7
		CCR2	CCL13
		CCR3	CCL11, CCL5, CCL7, CCL13

skin is an important immunological event in the development of allergic skin inflammation. There are higher proportions of CLA^+ $CCR4^+$ CD4 T cells circulating in the peripheral blood of patients with atopic dermatitis compared with unaffected individuals, and these cells are abundant in the skin lesions.

IL-4 and IL-13 also increase IgE synthesis by promoting B-cell isotype switching to IgE, and IL-4 stimulates the preferential differentiation of T_H2 cells from naive CD4 T cells after antigen encounter. IL-5 promotes the differentiation and survival of eosinophils, which secrete a range of inflammatory mediators. In addition, the T_H2 cytokine IL-31 induces itching, and Fas ligand and TNF-α expressed by activated T_H2 cells induce keratinocyte damage. T_H2 cytokines also downregulate the expression of antimicrobial peptides by keratinocytes, which may underlie the increased susceptibility of patients with atopic dermatitis to cutaneous bacterial and viral infections that include *S. aureus*, herpes simplex type 1, molluscum contagiosum, and vaccinia virus.

In addition to infiltration by T_H2 cells, there is also evidence of a modest infiltration of T_H17 and T_H22 cells, which secrete IL-17 and IL-22, respectively, although these cytokines are much more prominent in psoriasis, another inflammatory skin disease. In chronic atopic dermatitis lesions there is a mixture of infiltrating T_H1 and T_H2 cells. The reason for the switch from T_H2 cell-dominated infiltrates in acute atopic dermatitis lesions to a mixed infiltrate in chronic lesions is not known. Microbial products that act via Toll-like receptors such as TLR-9 may promote a T_H1 response to microbial antigens in chronic atopic dermatitis.

The Langerhans cells and macrophages that infiltrate the skin lesions have IgE bound to their surface through CD23, a low-affinity receptor for IgE. Langerhans cells and monocytes can present antigen to naive T cells and activate them, and the bound IgE on the surface of the antigen-presenting cells allows them to concentrate the antigen, rendering them more efficient at antigen presentation. IgE is also bound to mast cells in the tissues through the high-affinity IgE receptor FcεRI. Signaling through this receptor after the cross-linking of IgE by antigen leads to the production and secretion of IL-4 and IL-5 by the mast cells, thus further biasing the T-cell response to a T_H2 phenotype. In chronic atopic dermatitis, the dermis is infiltrated by dendritic cells that bear IgE bound to the high-affinity IgE receptor on their surface. These cells secrete large amounts of IL-12 upon IgE cross-linking and therefore may be critical in inducing infiltrating T cells to secrete IFN-γ, a hallmark of chronic atopic dermatitis lesions.

Treatment of atopic dermatitis is aimed at softening the underlying dry skin and reducing inflammation. Emollients are the first line of topical therapy for the condition because they improve the skin's barrier function. Avoidance of irritants such as soaps and synthetic fabrics, which disrupt this barrier, is crucial in controlling atopic dermatitis. Acute flare-ups require treatment with topical corticosteroid and/or calcineurin inhibitor ointments or creams. Antihistamines can be helpful, especially at night, to control the itching.

Questions.

1 *Topical steroids are effective in reducing the eczema associated with atopic dermatitis. Why?*

2 *What other immunomodulatory agents might be effective in atopic dermatitis?*

3 *Why did Tom develop an extensive herpesvirus infection?*

4 *Atopic dermatitis is described as the 'itch that rashes.' If patients are prevented from scratching, no rash occurs. What is the relationship of scratching to the rash?*

5 *Skin infections with staphylococci and other bacteria exacerbate atopic dermatitis. Can you suggest a possible explanation for this?*

6 *Many patients with atopic dermatitis have associated asthma and/or food allergy. Why?*

CASE 6 Contact Sensitivity to Poison Ivy

A delayed hypersensitivity reaction to a hapten.

Allergic or hypersensitivity reactions can be elicited by antigens that are not associated with infectious agents, for example pollen, dust, food, and chemicals in the environment. They do not usually occur on the first encounter with the antigen, but a second or subsequent exposure of a sensitized individual causes an allergic reaction. Allergic symptoms will depend on the type of antigen, the route by which it enters the body, and the cells involved in the immune response. These unwanted responses can cause distressing symptoms, tissue damage, and even death. These are the same reactions that would be provoked by a pathogenic antigen, had it been introduced and presented in the same way. When they are not helping to combat an infection, however, these damaging side-effects are clearly unwanted.

There are four main types of immunological hypersensitivity reactions, which are distinguished by the type of immune cells and antibodies involved, and the pathologies produced. The one discussed here is an example of a type IV (delayed hypersensitivity) reaction (Fig. 6.1). Many allergic reactions occur within minutes or a few hours of encounter with the antigen, but some take a day or two to appear (Fig. 6.2). The latter are the delayed hypersensitivity reactions. Delayed hypersensitivity reactions are mediated by T cells only, either T_H1 CD4 T cells or cytotoxic CD8 T cells, or sometimes both. Antibodies are not involved. The reactions can be triggered by foreign proteins or by self proteins that have become modified by the attachment of a hapten, such as a small organic molecule or metal ion. A common type of delayed hypersensitivity reaction is allergic contact dermatitis, a skin rash caused by direct contact with the antigen.

Delayed hypersensitivity reactions fall into two classes (see Fig. 6.1). In the first, the damage is due to an inflammatory response and tissue destruction by T_H1 cells and the macrophages they activate. In the second class of delayed

This case was prepared by Raif Geha, MD, in collaboration with Lisa Bartnikas, MD.

Topics bearing on this case:

- T-cell priming (sensitization)
- Apoptosis
- Preferential activation of T_H1 cells
- Allergic reactions
- Inflammatory response

Type IV immune-mediated tissue damage		
Immune reactant	T cells	
Antigen	Soluble antigen	Cell-associated antigen
Effector mechanism	Macrophage activation	Cytotoxicity
	T_H1 cytotoxins	CTL
Example of hypersensitivity reaction	Allergic contact dermatitis, graft rejection	
	Rheumatoid arthritis	Diabetes mellitus

Fig. 6.1 Type IV hypersensitivity reactions. There are four types of immune-mediated tissue damage. Types I–III are antibody-mediated and are distinguished by the different types of antigens recognized and the different classes of antibodies involved. We have seen examples of type I responses in Case 1 and Case 2 and will see type III in Case 8. Type IV hypersensitivity reactions are T cell-mediated and can be subdivided into two classes. In the first class, tissue damage is caused by T_H1 cells, which activate macrophages, leading to an inflammatory response. On encounter with antigen, effector T_H1 cells secrete cytokines, such as interferon-γ, that activate macrophages, and to a lesser extent mast cells, to release cytokines and inflammatory mediators that cause the symptoms. In the second class of type IV reactions, damage is caused directly by cytotoxic T lymphocytes (CTLs) that attack tissue cells presenting the sensitizing antigen on their surface. The delay in the appearance of a type IV hypersensitivity reaction is due to the time it takes to recruit antigen-specific T cells and other cells to the site of antigen localization and to develop the inflammatory response. Because a delayed hypersensitivity response involves antigen processing and presentation to achieve T-cell activation, quite large amounts of antigen need to be present at the site of contact. The amount of antigen required is two or three orders of magnitude greater than that required to initiate an antibody-mediated immediate hypersensitivity reaction.

hypersensitivity reactions, tissue damage is caused mainly by the direct action of antigen-specific cytotoxic CD8 T cells on target cells displaying the foreign antigen. Some antigens may cause a combination of both types of reactions.

This case describes the most frequently encountered delayed hypersensitivity reaction in the United States—allergic contact dermatitis due to the woodland plant poison ivy.

The case of Paul Stein: a sudden appearance of a severe rash.

Paul Stein was 7 years old and had enjoyed perfect health until 2 days after he returned from a hike with his summer camp group, when itchy red skin eruptions appeared all along his right arm. Within a day or two, the rash had spread to his trunk, face, and

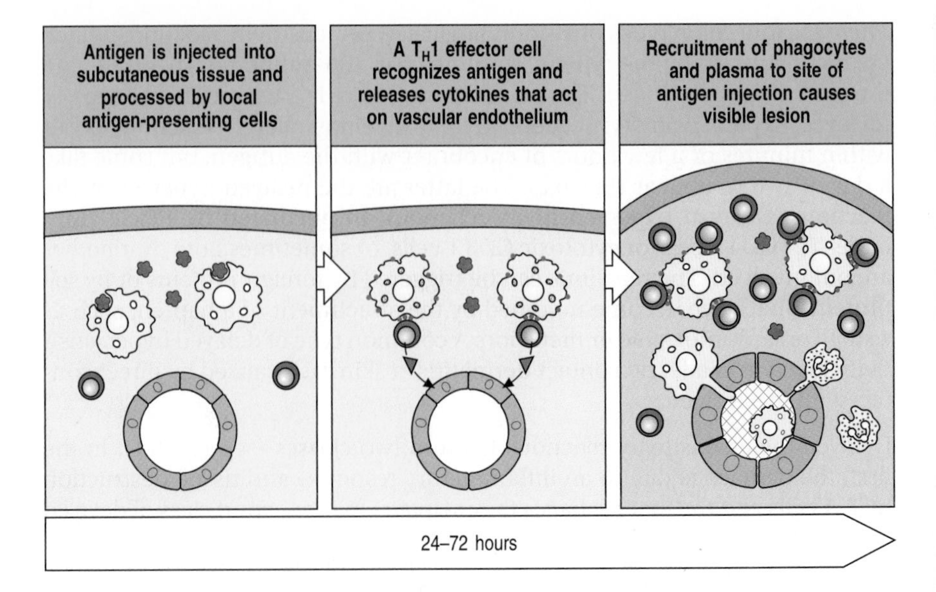
Antigen is injected into subcutaneous tissue and processed by local antigen-presenting cells
A T_H1 effector cell recognizes antigen and releases cytokines that act on vascular endothelium
Recruitment of phagocytes and plasma to site of antigen injection causes visible lesion
24–72 hours

Fig. 6.2 The time course of a delayed-type hypersensitivity reaction. The first phase involves the uptake, processing, and presentation of the antigen by local antigen-presenting cells. In the second phase, T_H1 cells that were primed by a previous exposure to the antigen migrate into the site of injection and become activated. Because these specific cells are rare, and because there is no inflammation to attract cells into the site, it may take several hours for a T cell of the correct specificity to arrive. These cells release mediators that activate local endothelial cells, recruiting an inflammatory cell infiltrate dominated by macrophages and causing the accumulation of fluid and protein. At this point, the lesion becomes apparent.

genitals. His mother gave him the antihistamine Benadryl (diphenhydramine hydrochloride) orally to suppress the itching, but this gave only partial relief. The rash did not improve, and a week after it first appeared he attended the Dermatology Clinic at the Children's Hospital.

Physical examination revealed large patches of raised, red, elongated blisters, oozing scant clear fluid, on his body and extremities (Fig. 6.3). Paul also had swollen eyelids and a swollen penis. There was no history of fever, fatigue, or any other symptom. A contact sensitivity reaction to poison ivy was diagnosed.

He was given a corticosteroid-containing cream to apply to the skin lesions three times a day, and Benadryl to take orally three times a day. He was asked to shampoo his hair, wash his body thoroughly with soap and water, and cut his nails short.

Two days later, his parents reported that, although no new eruptions had appeared, the old lesions were not significantly better. Paul was then given the corticosteroid prednisone orally, which was gradually decreased over a period of 2 weeks. The topical steroid cream was discontinued.

Within a week, the rash had almost disappeared. Upon stopping the prednisone there was a mild flare-up of some lesions, and this was controlled by application of topical steroid for a few days. Paul was shown how to identify poison ivy in order to avoid further contact with it, and told to wear long pants and shirts with long sleeves on any future hikes in the woods.

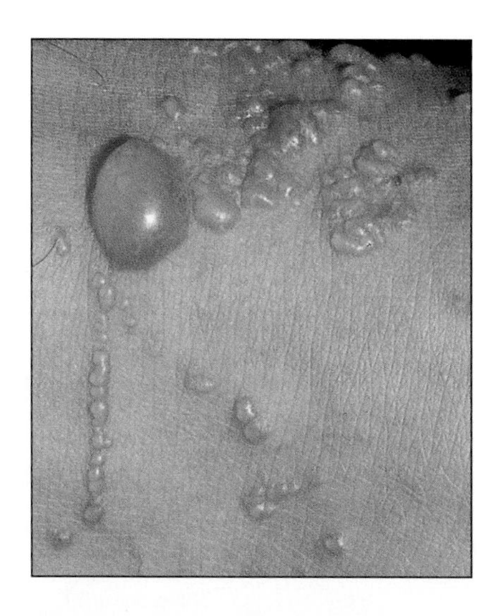

Fig. 6.3 Blistering skin lesions of patient with poison ivy contact dermatitis. Note the linear pattern of blisters in several areas. This is called the Koebner phenomenon and is due to exposure to the hapten along a line, possibly due to initial wiping or scratching. A rash only occurs on the initial areas of skin contact. Once the hapten is cleaned from the skin, no additional areas of skin will become involved. Therefore, touching skin lesions or blister fluid will not result in additional spread.

Contact sensitivity to poison ivy.

The reaction to poison ivy is the most commonly seen delayed hypersensitivity reaction in those parts of the United States where the plant grows wild. The absence of fever or general malaise accompanying the rash, and Paul's otherwise excellent health except for the skin lesions, point to a contact sensitivity reaction rather than to a viral or bacterial infection or some other underlying long-term illness. The appearance of the rash just 2 days after Paul returned from a hiking trip where he could easily have been in contact with poison ivy virtually clinches the diagnosis.

Allergic contact dermatitis due to poison ivy is caused by a T-cell response to a chemical in the leaf called pentadecacatechol (Fig. 6.4). On contact with the skin, this small, highly reactive, lipid-like molecule penetrates the outer layers, and binds covalently and nonspecifically to proteins on the surfaces of skin cells, in which form it functions as a hapten. Most people are susceptible, and sensitivity, once acquired, is lifelong.

The generation of a delayed-type hypersensitivity reaction requires the completion of both an 'afferent' and an 'efferent' response. In the afferent part of the response, the hapten enters the epidermis, and haptenated self proteins are ingested by specialized phagocytic cells in the epidermis (Langerhans cells) and dermis (dendritic cells) into intracellular vesicles, where they are cleaved into peptides. Some of these peptides will have hapten attached. The peptides bind to MHC class II molecules in the vesicles and are presented as peptide:MHC complexes on the Langerhans cell surface. Over the next 12–48 hours, some of these Langerhans cells and dermal dendritic cells migrate to a regional lymph node, where they become antigen-presenting cells that can activate naive hapten-specific T cells to become recirculating hapten-specific effector CD4 T_H1 cells and effector cytotoxic CD8 T cells that express skin-homing receptors such as E-selectin and CCR4. The efferent part of the response involves homing by these sensitized T_H1 and cytotoxic T cells to the site of contact with the plant. There, the activated effector T cell can

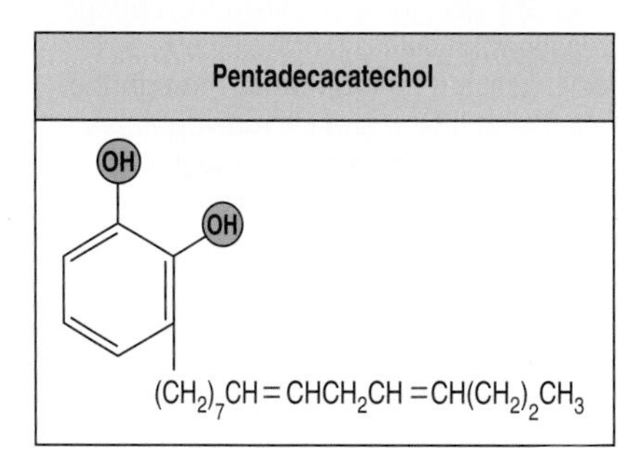

Fig. 6.4 The chemical formula of pentadecacatechol, the causative agent of contact sensitivity to poison ivy.

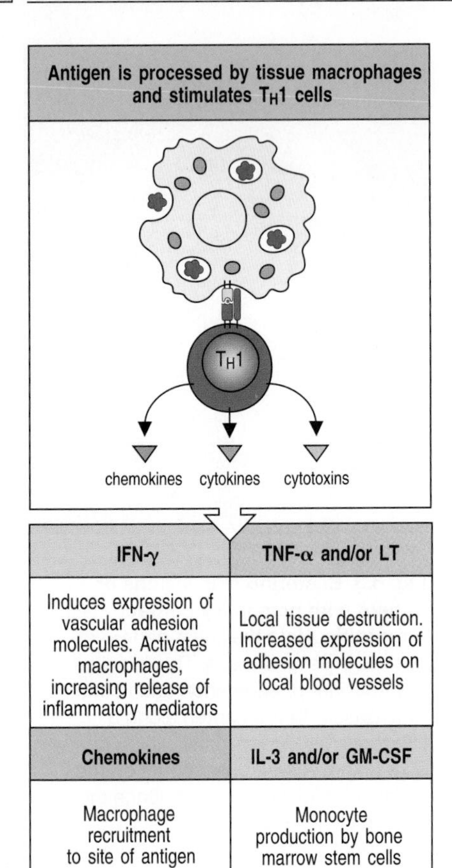

Antigen is processed by tissue macrophages and stimulates T_H1 cells
T_H1
chemokines
cytokines
cytotoxins
IFN-γ
TNF-α and/or LT
Induces expression of vascular adhesion molecules. Activates macrophages, increasing release of inflammatory mediators
Local tissue destruction. Increased expression of adhesion molecules on local blood vessels
Chemokines
IL-3 and/or GM-CSF
Macrophage recruitment to site of antigen
Monocyte production by bone marrow stem cells

Fig. 6.5 The delayed-type (type IV) hypersensitivity response is directed by cytokines released by T_H1 cells stimulated by antigen. Antigen in the local tissues is processed by antigen-presenting cells and presented on MHC class II molecules. Antigen-specific T_H1 cells can recognize the antigen locally at the site of injection, and release chemokines and cytokines that recruit macrophages to the site of antigen deposition. Antigen presentation by the newly recruited macrophages then amplifies the response. T cells may also affect local blood vessels through the release of TNF-α and LT (lymphotoxin) and stimulate the production of macrophages through the release of IL-3 and GM-CSF (granulocyte–macrophage colony-stimulating factor). Finally, T_H1 cells activate macrophages through the release of IFN-γ, and kill macrophages and other sensitive cells through the release of LT or by the expression of Fas ligand.

react with haptenated peptides presented by Langerhans cells and dermal dendritic cells, leading to a release of inflammatory mediators by the T_H1 cells and cytotoxic molecules, such as perforin, by the cytotoxic T cells, thereby initiating a local delayed hypersensitivity reaction in the skin. With each subsequent exposure to antigen, the period of latency from contact to appearance of a rash is shortened (anamnesis).

The appearance of Paul's rash 2 days after his suspected exposure to poison ivy is typical of a delayed hypersensitivity reaction. The haptenated self peptides presented on skin macrophages and Langerhans cells at the site of contact with poison ivy are initially recognized by the small number of activated hapten-specific T_H1 cells within the pool of recirculating T cells. On encounter with the haptenated peptides, these T_H1 cells release chemokines, cytokines, and cytotoxins that initiate an inflammatory reaction and also kill cells directly (Fig. 6.5).

One of the cytokines produced by the T_H1 cells is interferon-γ (IFN-γ), whose main effect in this context is to activate macrophages. The subsequent macrophage activity causes many of the symptoms of the delayed hypersensitivity reaction. Macrophages activated by IFN-γ release cytokines and inflammatory mediators such as interleukins, prostaglandins, nitric oxide (NO), and leukotrienes. The combined effects of T-cell and macrophage activity cause a local inflammatory response and tissue damage at the site of the contact with poison ivy.

The red, raised, blistering skin lesions of poison ivy dermatitis are due to the infiltration of large numbers of blood cells into the tissue at the site of contact, combined with the localized death of tissue cells and the destruction of the extracellular matrix that holds the layers of skin together. One of the first actions of effector T_H1 cells on contact with their antigen is to release the cytokines TNF-α and LT (lymphotoxin), and chemokines such as CCL5 (formerly known as RANTES). TNF-α in particular increases the expression of adhesion molecules on the endothelium lining postcapillary venules and increases vascular permeability so that macrophages and other leukocytes adhere to the sides of the blood vessel. This aids their migration from the bloodstream into the tissues in response to the secreted chemokines.

Once at the contact site, macrophages are activated and themselves release cytokines and other inflammatory mediators, which attract more monocytes, T cells, and other leukocytes to the site, thus helping to amplify and maintain the inflammatory reaction. The blood vessels also dilate, which causes the redness associated with the rash. The inflammatory mediators also act on mast cells to cause degranulation and the release of histamine, which is the main cause of the itching that accompanies the reaction.

Tissue destruction, which is a feature of delayed hypersensitivity reactions, is caused both by cytokines and by direct cell–cell interactions. The TNF-α and LT released by T_H1 cells and macrophages act at the same TNF receptors, which are expressed on virtually all types of cells, including skin cells. Stimulation of these receptors induces a 'suicide' pathway in the cells—apoptosis—which causes their death. Activated CD4 T cells also express the Fas ligand (FasL) in their plasma membrane, which interacts with the ubiquitously expressed cell-surface molecule Fas to cause the death of the target cell by apoptosis. Other mediators released by activated T cells, such as the enzyme stromelysin, degrade the proteins of the extracellular matrix, which maintain the integrity of the skin.

Lipid-like haptens such as pentadecacatechol can also cause the priming and activation of cytotoxic CD8 T cells, as small fat-soluble molecules can enter the cytosol of skin cells directly by diffusing through the plasma membrane.

Once inside, pentadecacatechol binds to intracellular proteins. Peptides generated from the haptenated proteins in the cytosol are delivered to the cell surface associated with MHC class I molecules. These target cells are recognized and attacked by antigen-specific cytotoxic CD8 T cells, which have become primed and activated on a previous encounter with the antigen. The outcome of all these reactions is the raised, red, weeping blisters characteristic of sensitivity to poison ivy.

Corticosteroids are the standard treatment for hapten-mediated contact sensitivity, because they inhibit the inflammatory response by inhibiting the production of many of the cytokines and chemokines. Corticosteroids are lipid-like molecules that can diffuse freely across plasma membranes. Once inside the cell, they bind to receptor proteins in the cytoplasm. The receptor:steroid complex enters the nucleus, where it controls the expression of several genes. Of relevance here is the fact that it induces the production of an inhibitor of the transcription factors required to switch on transcription of the cytokine and chemokine genes. In mild cases of poison ivy dermatitis, topical steroids applied locally are sufficient. In more severe cases such as Paul's, oral steroids are needed to achieve a concentration necessary to inhibit the inflammatory response.

Paul was given antihistamines to block the histamine receptors and counteract the action of the histamine released from mast cells. Antihistamines also counteract itching caused by substances other than histamine (for example, prostaglandins released from macrophages).

Questions.

1 *Paul had lesions not only on the exposed skin but also on areas that would be covered, like the trunk and penis and in areas that were not in obvious contact with poison ivy leaves. How do you explain that?*

2 *How do you explain the recurrence of the lesions after discontinuation of the corticosteroids?*

3 *Paul must take great care to avoid poison ivy all his life, because subsequent reactions to it could be even more severe. Why would this be?*

4 *How would you confirm that Paul's contact dermatitis was caused by poison ivy rather than by another chemical such as the one found in the leaves of poison sumac (*Toxicodendron vernix*), another plant that gives rise to contact dermatitis?*

5 *One of Paul's friends, Brian, has X-linked agammaglobulinemia. What is the likelihood that this boy will develop poison ivy sensitivity?*

6 *Delayed hypersensitivity reactions are a rapid, inexpensive, and easy measure of T-cell function. What antigens might you use to test people for T-cell function in this way?*

7 *What are some common causes of contact sensitivity?*

CASE 7 Transfusion Reaction in IgA Deficiency

An anaphylactic reaction to a platelet transfusion uncovers an undiagnosed immune deficiency.

Transfusion of red cells and other blood components is used both to compensate for general blood loss due to trauma or surgery, and to replace missing blood components in the management of a number of hematologic disorders. Plasma and blood cells contain a large pool of potential antigens, many of which differ between individuals, and immune sensitization of the patient to antigens in the donated blood can lead to complications during transfusion.

For use in transfusion, whole blood is fractionated into three different components: red cells, platelets, and plasma, which are used for different purposes (Fig. 7.1). Blood transfusions comprising red cells only require matching for the red blood cell antigens of the ABO and Rh systems. Such matching is needed because, depending on their blood type, people have preexisting 'natural antibodies' that will cross-react with blood group antigens of a different type than their own. Natural antibodies arise in the course of normal immunologic development even in the absence of exposure to the relevant antigens. Typically, they are low-affinity, highly cross-reactive antibodies that are generated in response to oligosaccharides present on bacteria colonizing the intestine. If mismatched blood is transfused, the preexisting natural antibodies in the donor bind to the incoming red cells, leading to hemolysis.

Red cells, blood plasma, and platelets carry a variety of other antigens that can provoke immune reactions. Antibodies against these can be generated in response to direct exposure, in a conventional adaptive immune response, such as might occur in the setting of repeated blood transfusions or during pregnancy in a mother whose fetus expresses blood cell antigens different from her own.

The cause and clinical presentation of transfusion reactions vary widely, and depend on the particular blood component transfused and the immune status of the patient. For example, plasma does not contain cells or platelets but does contain a significant amount of donor immunoglobulin, and so has the potential to react with antigens in the recipient's tissue. Some cases of transfusion reaction may even be nonimmunologic, being induced directly by mediators of hypersensitivity, such as histamine or cytokines, generated within the transfused product. For example, preparations of platelets, which are typically stored in the blood bank at room temperature rather than being refrigerated, can accumulate reactive cytokines and histamine generated by the platelets themselves. For all these reasons, each case of a transfusion reaction must be carefully assessed by considering the constituents of the blood component being transfused.

Topics bearing on this case:
Natural antibodies
IgA deficiency
Common variable immunodeficiency
IgE-mediated allergic reactions
Blood groups
Indirect and direct antiglobulin tests

This case was prepared by Hans Oettgen, MD, PhD, and Raif Geha, MD, in collaboration with John Manis, MD.

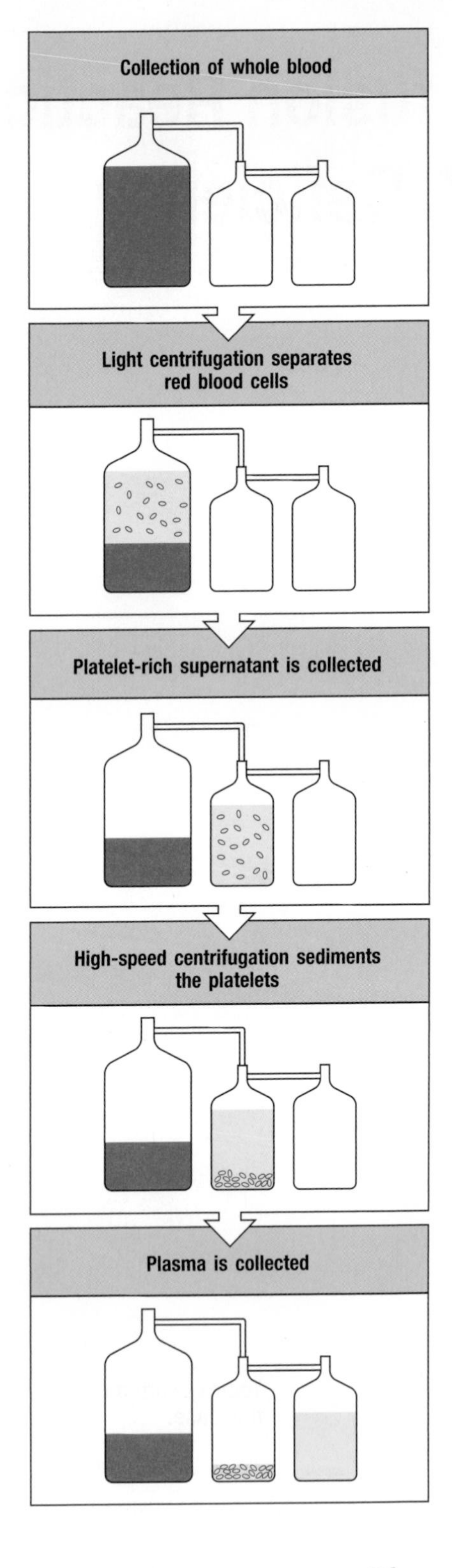

Platelet-rich supernatant is collected
High-speed centrifugation sediments the platelets
Plasma is collected

Fig. 7.1 Blood components used for transfusion. For transfusion, whole blood is fractionated into three components: red cells, platelets, and plasma. White cells are removed because they carry HLA antigens, and HLA matching is impracticable in this context. After collection, blood is first centrifuged to separate out the red cells, which are stored in specialized solutions containing relatively little plasma. The platelet-rich plasma fraction is then spun to separate plasma (which is usually immediately frozen) from platelets. Platelets are stored in small amounts of plasma at room temperature on a shaking platform. Platelets may also be collected directly from donors with an apheresis instrument, which automatically collects platelets together with plasma and returns red blood cells to the donor.

In the Case discussed here, a transfusion of platelets to correct a platelet deficiency provoked an immediate IgE-mediated allergic reaction in the recipient, even though donor and recipient had been matched for the ABO blood group antigens. As we shall see, an underlying immune deficiency puts this particular patient at a higher than usual risk of such a reaction.

The case of Nathan Zuckerman: a 25-year-old student with a severe nosebleed.

Nathan was referred to the Children's Hospital Emergency Department for a severe nosebleed. It had come on suddenly while he was studying and would not stop, even though he applied steady pressure to his nose. Nathan told the doctor that over the course of the past two weeks tiny red spots had appeared on his skin, as well as bruises over his arms, thighs, and abdomen.

Nathan's past medical history was remarkable for recurrent and persistent infections. These included otitis media, which had required a tympanostomy tube, two bacterial pneumonias, and frequent episodes of sinusitis. Most of these illnesses responded to antibiotics and Nathan had never required hospitalization. His younger brother had a similar problem with frequent infections, and both the brothers developed more frequent episodes when they became young adults. Nathan had no history of allergy or asthma.

On physical examination in the Emergency Department, Nathan's vital signs were within the normal range and he had no fever. Scattered petechiae (tiny red spots that did not whiten under pressure) were noted on his skin, and bruises of various ages and sizes were observed on his trunk and extremities. The combination of persistent nosebleed, bruises, and petechiae raised a strong suspicion of platelet dysfunction.

Nathan's nose was packed with gauze, which rapidly became saturated with blood. Blood samples were sent to the laboratory for a complete blood count with differential (CBC) as well as tests for blood type and an indirect antiglobulin test for anti-red-cell antibodies (antibody screen) (Fig. 7.2). The CBC revealed thrombocytopenia (platelet deficiency) with a platelet count of 9×10^9 l^{-1} (normal 140–450 $\times 10^9$ l^{-1}), but normal levels of white cells (7.8×10^9 l^{-1}) and red cells (as determined by hemoglobin levels (124 g l^{-1})).

25-year-old man
with severe nosebleed,
bruises, and petechiae.

An otolaryngologist was consulted for further management of the nosebleed, and a platelet transfusion was ordered. Nathan had never previously been transfused, and the platelets that were obtained matched his blood type. Ten minutes after starting the infusion, having received only 20 ml of platelets, he developed urticaria, but no other symptoms. The transfusion was paused and Nathan was given 25 mg of the histamine-receptor blocker diphenhydramine intravenously. This ameliorated the

urticaria. The platelet transfusion was about to be resumed when Nathan suddenly developed breathing difficulty (dyspnea), wheezing, and nausea. He was given oxygen, more diphenhydramine, and intravenous corticosteroids, but showed no improvement. Within minutes he became hypotensive, developed stridor and became hypoxic, despite being given supplementary oxygen. The otolaryngologist intubated Nathan to protect his airway and noted significant laryngeal edema while placing the endotracheal tube. Epinephrine (adrenaline), diphenhydramine, and fluids were given and Nathan was transferred to the intensive care unit, where he gradually recovered over the next 48 hours. A repeat CBC was obtained immediately after the transfusion, as was a repeat antibody screen and a direct antiglobulin test for red-cell-bound antibodies (see Fig. 7.2). There were no significant changes in Nathan's blood cell count (white blood cells: 15.6 × 10^9 l^{-1}; platelets: 119 g l^{-1}), he had a normal lactate dehydrogenase level (which would have been elevated if intravascular hemolysis had been occurring), and both the antibody screen and a direct antiglobulin test for IgG bound to his red blood cells were negative.

Thrombocytopenia with normal red blood cell and white blood cell levels.

Systemic anaphylaxis following platelet transfusion.

Transfusion reaction in IgA deficiency.

Allergic reactions are the most common adverse responses encountered when administering blood products. They are typically precipitated by transfusions of platelets or plasma, and are characterized by localized or generalized urticaria, erythema, and pruritus. Unlike Nathan's reaction of systemic anaphylaxis, most allergic transfusion reactions are mild, transient, and unaccompanied by significant systemic symptoms. Allergic transfusion reactions are triggered by IgE or IgG antibodies in the recipient directed against cell proteins or soluble allergens in the transfused blood product.

The overall incidence of mild urticarial reactions may be as high as 1–3% of all plasma-containing transfusions. Most mild reactions disappear shortly after stopping the transfusion, but urticaria can be treated with histamine-receptor blockers if needed. Red cells or platelets can be washed to remove plasma containing the causative factors before subsequent infusions in individuals who have experienced an allergic reaction.

Anaphylactic transfusion reactions, as seen in Nathan, are severe systemic type I hypersensitivity responses characterized by hypotension, bronchospasm, dyspnea, angioedema, nausea, vomiting, diarrhea, and urticaria (see Case 1). Laryngeal edema and hypotension are the most serious, potentially fatal, complications. Anaphylaxis is rare during blood transfusion, occurring at a frequency of 1 in 20,000 to 1 in 50,000 of all blood transfusions, with one or two fatalities per year reported in the United States. The clinical hallmark of anaphylactic transfusion reactions is that they are rapid in onset and are seen after as little as 10 ml of product has been transfused. Fever is not typically observed.

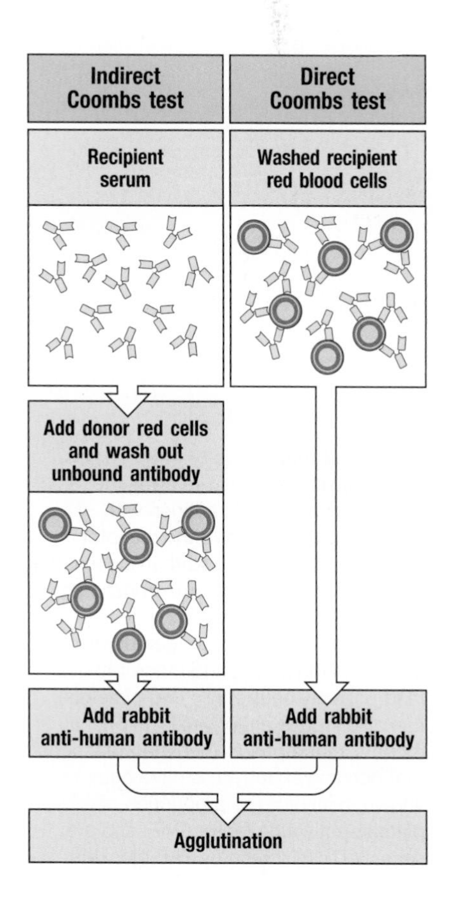

Fig. 7.2 Indirect and direct antiglobulin tests. These tests are used for the detection of antibodies against surface antigens on red blood cells. In the indirect antiglobulin test, washed red blood cells from the donor are incubated with the recipient's serum. Anti-human immunoglobulin antibody is then added to detect any recipient antibody that has bound to the donor red cells. The direct antiglobulin test detects donor antibodies bound to the recipient's red cells after transfusion. Washed blood cells from the recipient are incubated with an anti-human immunoglobulin antibody. If agglutination (clumping) occurs, the test is positive.

Anaphylaxis during transfusion is mediated by IgE or IgG antibodies. In recipients with IgE antibodies directed against blood components, the response arises via the conventional allergic pathway (see Case 1) in which antigen-mediated aggregation of allergen-specific IgE bound to high-affinity FcεRI receptors on mast cells and basophils leads to activation of these cells (see Fig. 2.2). The resulting release of vasoactive mediators, including histamine and cysteinyl leukotrienes (see Fig. 2.7), leads to systemic vasodilation, plasma extravasation, and shock.

IgG antibodies can drive anaphylactic reactions by binding to Fcγ receptors (FcγRIII) on mast cells, macrophages, and basophils, activating them and leading to the release of vasoactive mediators. Unlike FcεRI, which has a very high affinity for its ligand, IgE, and is fully occupied under baseline physiologic conditions, FcγRIII has relatively low affinity for IgG and can interact only with polyvalent immune complexes of antigen and IgG that contain multiple Fcγ domains (Fig. 7.3). Thus, larger amounts of antigen, as well as relatively high levels of specific IgG antibodies, are necessary to trigger IgG-mediated anaphylaxis compared with the IgE-mediated reaction. FcγRIII-activated macrophages release platelet-activating factor, which has vascular effects and also further activates mast cells. In transfusion-mediated anaphylaxis, IgE- and IgG-mediated mechanisms may overlap.

Diagnostically, anaphylaxis during a transfusion reaction must be differentiated from more common acute reactions, including transfusion-related acute lung injury (TRALI) (Fig. 7.4). TRALI is characterized by fever, tachycardia, and tachypnea (increased respiratory rate), usually occurring within 6 hours of transfusion. The pathogenesis of TRALI is not completely understood, but it is known to be associated with the infusion of donor anti-leukocyte antibodies (for example, anti-HLA) that can interact with primed neutrophils in the recipient, leading to their activation and aggregation in the pulmonary vasculature. TRALI occurs in about 1 in 5000 to 1 in 10,000 transfusions and is the leading cause of transfusion-associated deaths in the United States. For this reason, plasma from women, which may contain anti-HLA antibodies as a result of sensitization by fetal alloantigens during pregnancy, is not used for transfusion in the United States.

One of the best-documented causes of anaphylactic transfusion reactions is an anti-IgA reaction first seen in severely IgA-deficient patients, some of whom have anti-IgA antibodies in their serum. These anti-IgA antibodies react against any IgA in the transfusion product. The first cases were described more than 40 years ago, and since then IgA-driven transfusion reactions have been identified in people with other immunodeficiencies involving low or absent IgA, including common variable antigen deficiency (CVID) (Fig. 7.5).

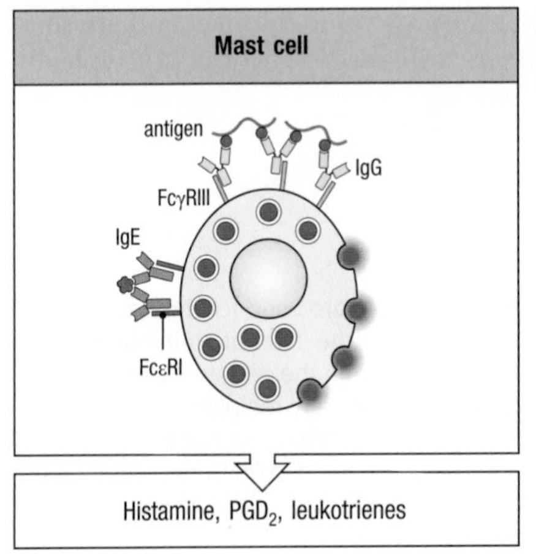

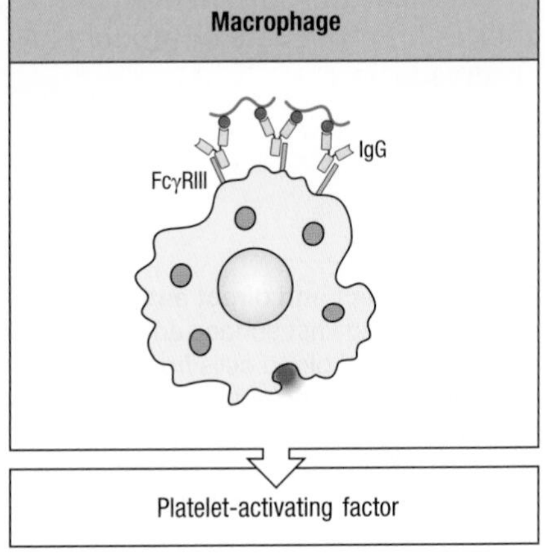

Fig. 7.3 Mechanisms of anaphylaxis. Anaphylaxis classically involves the interaction of specific antigen with preformed IgE bound to the receptor FcεRI on mast cells. In Nathan's case, the antigen is IgA in the plasma transfused along with the platelets. Immune complexes of IgG and antigen binding to FcγRIII on mast cells and macrophages (and possibly neutrophils and platelets) can also induce anaphylaxis. In animal models, macrophage-mediated reactions that occur in anaphylaxis have been shown to involve the production of platelet-activating factor (PAF) and are sensitive to inhibition by PAF blockers.

In Nathan's case, further investigations showed that an anti-IgA reaction was the probable cause of his reaction to the platelet transfusion.

Several clues in Nathan's case history hint at this diagnosis. His medical history was suggestive of an immune deficiency, possibly CVID because the bacterial infections became more frequent when he was at the age to enter college. CVID is one of the most prevalent primary immune deficiencies worldwide and is characterized by recurrent sinopulmonary infections presenting in the second and third decades of life. Individuals with CVID have defects in the production of antigen-specific IgG, and some also have decreased levels of IgA and/or IgM.

Patients with CVID often develop autoimmune diseases such as idiopathic thrombocytopenic purpurea (ITP), in which platelet destruction is mediated by autoantibodies directed against surface antigens on platelets and megakaryocytes. Nathan's initial blood count revealed a very low platelet count that was, given his normal white blood cell count and hematocrit, consistent with a diagnosis of ITP. Another clue to an underlying genetic immunodeficiency in Nathan was that his brother also suffered from frequent infections.

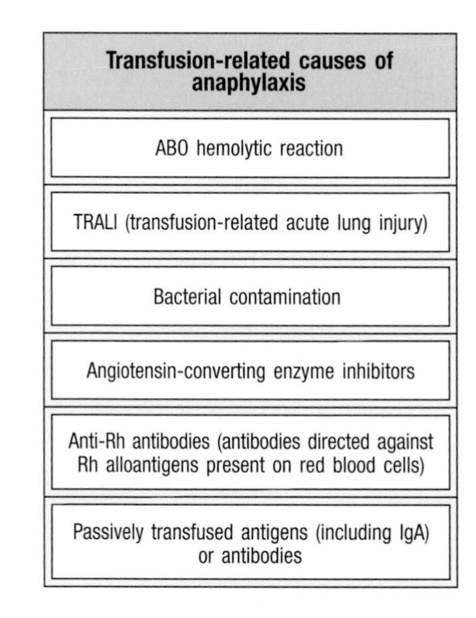

Transfusion-related causes of anaphylaxis
ABO hemolytic reaction
TRALI (transfusion-related acute lung injury)
Bacterial contamination
Angiotensin-converting enzyme inhibitors
Anti-Rh antibodies (antibodies directed against Rh alloantigens present on red blood cells)
Passively transfused antigens (including IgA) or antibodies

Fig. 7.4 Transfusion-related causes of anaphylaxis.

There are likely to be various genetic causes of CVID, most of them remaining to be discovered. The discovery of CVID-associated mutations in the *TACI* gene is of interest in the context of this Case, because patients with these mutations have low to no IgA. The TACI protein is required for efficient class-switch recombination, and it cooperates with signals induced by CD40 or by Toll-like receptors to induce immunoglobulin secretion and terminal B-cell differentiation. Notably, patients with CVID due to *TACI* mutations have a higher incidence of autoimmune and lymphoproliferative diseases. Subsequent tests determined that both Nathan and his brother have a mutation in the *TACI* gene and have concomitant low serum IgG and no IgA. Thus, taking a careful medical history may sometimes uncover patients at greater risk of anti-IgA-mediated transfusion reactions.

Selective IgA deficiency is the most common primary immunodeficiency in the Western world, with a prevalence ranging from 1 in 300 to 1 in 800 in those of European ethnicity compared with 1 in 5000 and 1 in 20,000 in China and Japan, respectively.

Patients lacking plasma IgA are the most likely to form antibodies against this isotype, either naturally or after exposure to IgA in plasma or intravenous

Clinical manifestations of severe IgA deficiency	
Recurrent sinopulmonary infections	Bronchitis (leading to bronchiectasis), pneumonia
Other infections	Meningococcemia, septic arthritis, conjunctivitis
Gastrointestinal tract disorders and infections	Inflammatory bowel disease, primary biliary cirrhosis, *Giardia lamblia, Helicobacter pylori*
Autoimmune disorders	Celiac disease, chronic nephritis, Evans syndrome, Hashimoto's thyroiditis, hemolytic anemia, idiopathic thrombocytopenic purpura, juvenile arthritis, myasthenia gravis, pernicious anemia, sarcoidosis, Sjögren's syndrome, systemic lupus erythematosus
Allergic reactions	Allergic rhinitis, atopic dermatitis
Malignancies	B-cell lymphoma, gastric cancer, multiple myeloma, melanoma

Fig. 7.5 Clinical manifestations of severe IgA deficiency. Patients with IgA most commonly present with a history of recurrent sinopulmonary infections, but can additionally be affected by a variety of other infectious and non-infectious disorders.

immunoglobulin. Severe IgA deficiency is defined by IgA levels of less than 0.05 mg dl^{-1}, and 20–35% of individuals in this group have anti-IgA antibodies. In one study of healthy blood donors in the United States, the prevalence of IgA deficiency was 0.26% and that of IgA deficiency with anti-IgA was 0.08% (about 1 in 1200). The frequency of anaphylactic transfusion reactions is far lower than would be predicted by this relatively high prevalence of anti-IgA antibodies, lending support to the hypothesis that additional factors are at play in patients with anti-IgA-related anaphylaxis. Nevertheless, patients with severe IgA deficiency and transfusion-associated anaphylaxis are more likely to have anti-IgA antibodies than similar patients who had no reactions. Thus it seems that the risk of anaphylaxis in patients with IgA deficiency is low, but that the presence of anti-IgA antibodies increases this risk. Most of the anti-IgA antibodies demonstrated have been of the IgG class, although IgE anti-IgA antibodies have been reported (see Fig. 7.4). Treatment of anti-IgA-mediated anaphylaxis is similar to that for anaphylaxis due to other causes, with supportive measures to support blood pressure and oxygenation.

Nathan presented with severe thrombocytopenia and significant bleeding. Given the urgency of his symptoms, a platelet transfusion was certainly required to stop his bleeding and to prevent life-threatening bleeds at other sites, including the brain. At present there are no entirely reliable and rapid assays available to detect anti-IgA antibodies in potential recipients of blood products. For this reason, screening strategies focus first on identifying patients with severe IgA deficiency and then on finding those with anti-IgA antibodies. Passive hemagglutination is one of the oldest and most widely used methods to detect anti-IgA antibodies, and is essentially the same assay as the indirect antiglobulin test (see Fig. 7.2). Indicator red blood cells are coated with IgA and incubated with serum (from either donors or patients). Clumping (agglutination) of the red blood cells indicates the presence of anti-IgA antibodies. This assay is relatively straightforward for a blood bank reference laboratory. Its sensitivity and specificity are low, however, and IgE anti-IgA antibodies cannot be detected. These limitations are driving efforts to develop more sensitive ELISA or particle gel immunoassays.

The management of blood product transfusion in patients with severe IgA deficiency or those with a documented history of anaphylactic reaction to IgA can be challenging. In severely IgA-deficient patients without a previous history of reaction, any available blood product can be administered, preferably in a clinical setting where anaphylaxis can be easily managed. Because not all patients with severe IgA deficiency and anti-IgA antibodies develop anaphylaxis, most individuals in this group can probably safely receive IgA-containing products. Once an IgA-mediated reaction has been demonstrated in a patient, specially prepared blood components such as washed red cells or washed platelets can be used for transfusion. If plasma is needed, this can be obtained from a registry that has products from tested IgA-deficient donors. In urgent clinical situations, however, there may not be enough time to find such a source, so clinicians must balance the risks and benefits of transfusion.

Questions.

1 *Apart from the presence of IgA deficiency and anti-IgA antibodies, what other diagnoses could explain anaphylaxis resulting from a platelet transfusion?*

2 *What would you do if Nathan needed to be transfused again?*

3 *Idiopathic thrombocytopenic purpura can be treated with intravenous immunoglobulin (IVIG), which at high doses can prevent platelet destruction by Fcγ receptor-dependent mechanisms. Could someone like Nathan have been treated more safely this way than with a platelet transfusion?*

4 *Do you think all patients with severe IgA deficiency should be screened for the presence of anti-IgA antibodies even if they do not require a transfusion? Give the reasons for your answer.*

5 *The most common cause of fatality associated with blood transfusions is TRALI. Can you suggest another cause of transfusion-associated fatality?*

CASE 8 Drug-Induced Serum Sickness

An adverse immune reaction to an antibiotic.

The intravenous administration of a large dose of antigen can evoke in some individuals a type III hypersensitivity reaction or immune-complex disease (Fig. 8.1). Antigen administration produces a rapid IgG response and the formation of antigen:antibody complexes (immune complexes) that can activate complement.

As a result of the large amount of antigen present and the rapid IgG response, small immune complexes begin to be formed in conditions of antigen excess (Fig. 8.2). Unlike the large immune complexes that are formed in conditions of antibody excess, which are rapidly ingested by phagocytic cells and cleared from the system, the smaller immune complexes are taken up by endothelial cells in various parts of the body and become deposited in tissues. Local activation of the complement system by these immune complexes provokes localized inflammatory responses.

The experimental model for immune-complex disease is the Arthus reaction, in which the subcutaneous injection of large doses of antigen evokes a brisk IgG response. The activation of complement by the IgG:antigen complexes generates the complement component C3a, a potent stimulator of histamine release from mast cells, and C5a, one of the most active chemokines produced by the body. The local endothelial cells are activated by the interactions in blood vessels between the immune complexes, complement and circulating leukocytes and platelets. They upregulate their expression of adhesion molecules such as selectins and integrins, which facilitates the emigration of white blood cells from the bloodstream and the initiation of a local inflammatory reaction (Fig. 8.3). Platelets also accumulate at the site, causing blood clotting; the small blood vessels become plugged with clots and burst, producing hemorrhage in the skin (Fig. 8.4).

When an antigen is injected intravenously, the immune complexes formed can be deposited at a wide range of sites. When deposited in synovial tissue, the resulting inflammation of the joints produces arthritis; in the kidney glomeruli they cause glomerulonephritis; and in the endothelium of the blood vessels of the skin and other organs they provoke vasculitis (Fig. 8.5).

This case was prepared by Raif Geha, MD.

Topics bearing on this case:

- Inflammatory reactions
- Properties of IgG antibodies
- Activation of complement by antigen:antibody complexes
- Type III immunological hypersensitivity reactions
- Immune-complex formation

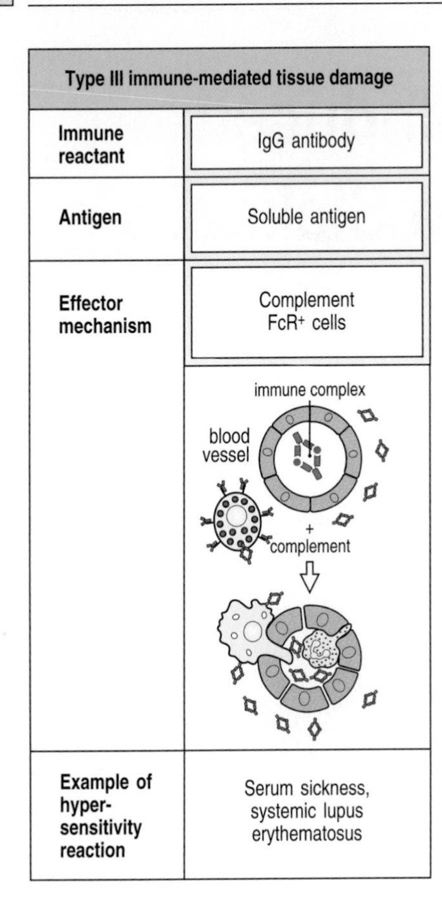

Type III immune-mediated tissue damage
Immune reactant
IgG antibody
Antigen
Soluble antigen
Effector mechanism
Complement
FcR+ cells
immune complex
blood vessel
complement
Example of hyper-sensitivity reaction
Serum sickness, systemic lupus erythematosus

Fig. 8.1 Type III hypersensitivity reactions. These can be caused by large intravenous doses of soluble antigens (serum sickness) or by an autoimmune reaction against some types of self antigen (as in systemic lupus erythematosus). The IgG antibodies produced form small immune complexes with the antigen in excess. The tissue damage involved is caused by complement activation and the subsequent inflammatory responses, which are triggered by immune complexes deposited in tissues.

In the early years of the twentieth century, the most common cause of immune-complex disease was the administration of horse serum, which was used as a source of antibodies to treat infectious diseases, and so this type of hypersensitivity reaction to large doses of intravenous antigen is still known as serum sickness.

This case describes a 12-year-old boy who received massive intravenous injections of penicillin and of ampicillin (one of its analogues) to treat pneumonia. He developed a serum-sickness reaction to the antibiotics.

The case of Gregory Barnes: serum sickness precipitated by penicillin.

When Gregory was brought to the Children's Hospital Emergency Room, his parents told the physicians that for 2 days he had had high fever (more than 39.5°C), a cough, and shortness of breath. Before then he had enjoyed excellent health. On physical examination he was pale, looked dehydrated, and was breathing rapidly with flaring nostrils. His respiratory rate was 62 min^{-1} (normal 20 min^{-1}), his pulse was 120 beats min^{-1} (normal 60–80 beats min^{-1}), and his blood pressure was 90/60 mmHg (normal). When his chest was examined with a stethoscope the emergency room doctors heard crackles (bubbly sounds) over the lower left lobe of his lungs. A chest radiograph revealed an opaque area over the entire lower lobe of the left lung. A diagnosis of lobar pneumonia was made.

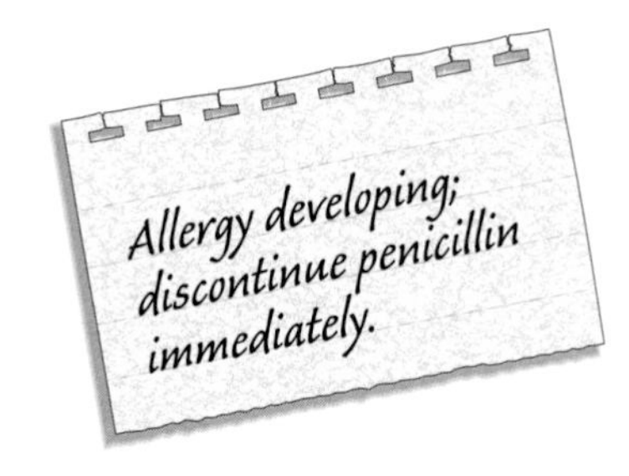

Allergy developing; discontinue penicillin immediately.

A white blood count revealed 19,000 cells μl^{-1} (normal 4000–7000 cells μl^{-1}) with an increase in the percentage of neutrophils to 87% of total white blood cells (normal 60%) and the abnormal presence of immature forms of neutrophils. A Gram stain of Gregory's sputum revealed Gram-positive cocci. Sputum and blood cultures grew *Streptococcus pneumoniae* (the pneumococcus).

Gregory was admitted to the hospital and treated with intravenous ampicillin at a dose of 1 g every 6 hours. Gregory gave no history of allergy to penicillin so ampicillin was used, to cover both Gram-positive and Gram-negative bacteria. On the fourth day of treatment, he felt remarkably better, his respiratory rate had decreased to 40 min^{-1}, and his temperature was 37.5°C. His white cell count had decreased to 9000 μl^{-1}. Because the *S. pneumoniae* grown from his sputum and blood was sensitive to penicillin, the ampicillin was replaced by penicillin. On his ninth day in hospital, Gregory had no fever, his white cell count was 7000 μl^{-1}, and his chest radiograph had improved. Plans for discharge from hospital were made for the following day.

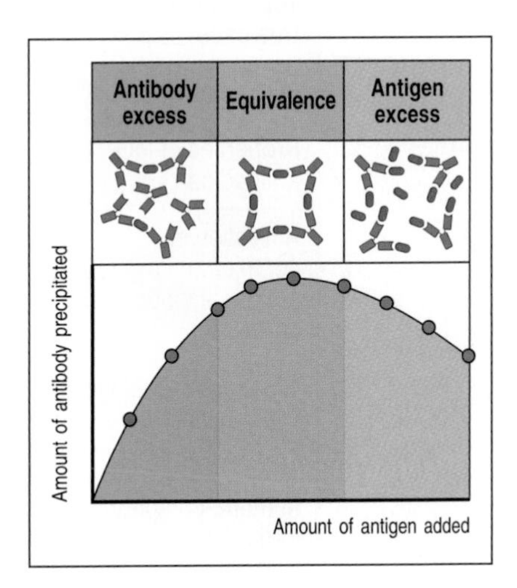

Fig. 8.2 Antibody can precipitate soluble antigen in the form of immune complexes. *In vitro*, the precipitation of immune complexes formed by antibody cross-linking the antigen molecules can be measured and used to define zones of antibody excess, equivalence, and antigen excess. In the zone of antigen excess, some immune complexes are too small to precipitate. When this happens *in vivo*, such soluble immune complexes can produce pathological damage to blood vessels.

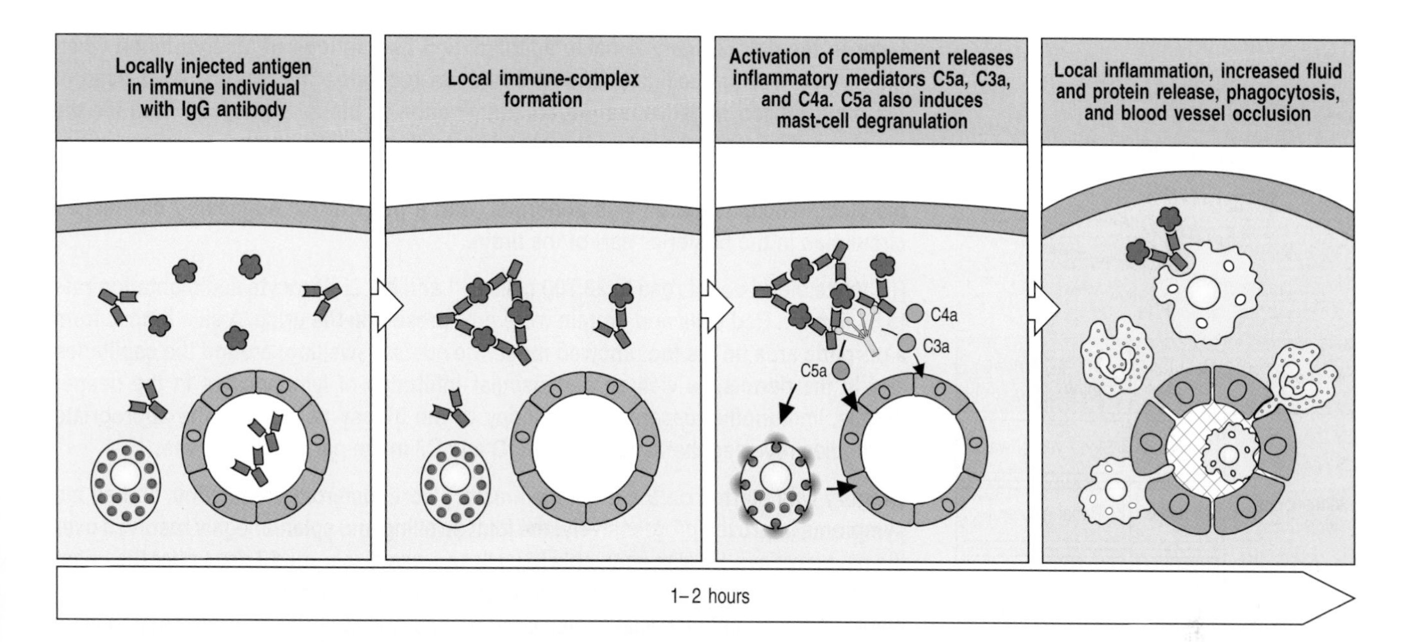

Locally injected antigen in immune individual with IgG antibody
Local immune-complex formation
Activation of complement releases inflammatory mediators C5a, C3a, and C4a. C5a also induces mast-cell degranulation
Local inflammation, increased fluid and protein release, phagocytosis, and blood vessel occlusion
C4a
C3a
C5a
1–2 hours

Fig. 8.3 The deposition of immune complexes in local tissues causes a local inflammatory response known as an Arthus reaction. In individuals who have already made IgG antibodies against an allergen, the same allergen injected into the skin forms immune complexes with IgG antibody that has diffused out of the capillaries. Because the dose of antigen is low, the immune complexes are only formed close to the site of injection, where they activate complement, releasing inflammatory mediators such as C5a, which in turn can activate mast cells to release inflammatory mediators. As a result inflammatory cells invade the site, and blood vessel permeability and blood flow are increased. Platelets also accumulate at the site, ultimately leading to occlusion of the small blood vessels, hemorrhage, and the appearance of purpura.

The next morning, Gregory had puffy eyes, and welts resembling large hives on his abdomen. He was given the antihistamine Benadryl (diphenhydramine hydrochloride) orally, and penicillin was discontinued. Two hours later he developed a tight feeling in the throat, a swollen face, and widespread urticaria (hives). With a stethoscope, wheezing could be heard all over his lungs. The wheezing responded to inhalation of the β_2-adrenergic agent albuterol. That evening Gregory developed a fever (a temperature of 39°C) and swollen and painful ankles, and his urticarial rash became more generalized. He appeared once again acutely ill.

The rash spread over his trunk, back, neck, and face, and in places became confluent (Fig. 8.6). Gregory also had reddened eyes owing to inflamed conjunctivae, and had swelling around the mouth. The anterior cervical, axillary, and inguinal lymph nodes on both sides were enlarged, measuring 2 cm by 1 cm. The spleen was also enlarged, with its tip palpable 3 cm below the rib margin. Ankles and knee joints were swollen and tender to palpation, and were too painful to move very far. The child was alert and his neurologic examination was normal.

Laboratory analysis of a blood sample revealed a raised white blood cell count (19,800 μl^{-1}) in which the predominant cells were lymphocytes (72%, in contrast with the normal 30%). Plasma cells were detected in a blood smear, although plasma cells are normally not present in blood. The erythrocyte sedimentation rate, an indicator of the presence of acute-phase reactants in the blood, was elevated at 30 mm h^{-1} (normal less than 20 mm h^{-1}). His total serum complement level and his serum C1q and C3 levels were decreased.

A presumptive diagnosis of serum sickness was made, and Gregory was given Benadryl and Naprosyn (naproxen), a nonsteroidal anti-inflammatory agent. On the following day, the rash and joint swellings were worse and the child complained of abdominal pain. There were also purpuric lesions, caused by hemorrhaging of small blood vessels under the skin, on his feet and around his ankles. There was no blood in his stool.

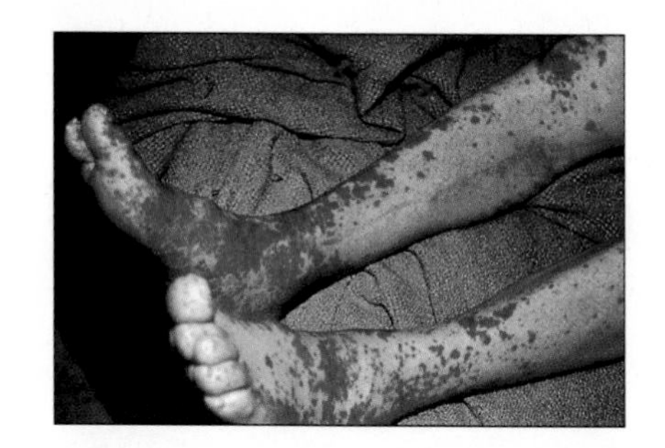

Fig. 8.4 Hemorrhaging of the skin in the course of a serum-sickness reaction.

Serum complement
components low.
Serum sickness.

Later in the day, Gregory became agitated, and had periods of disorientation when his speech was unintelligible and he could not recognize his parents. A CT scan of his brain proved negative, as did an examination of his cerebrospinal fluid for the presence of inflammatory cells, increased protein concentration, and decreased sugar concentration, all of which are indicators of infection and inflammation. However, his electroencephalogram was abnormal, with a pattern that suggested diminished circulation in the posterior part of the brain.

His white blood count rose to 23,700 cells μl^{-1} and his erythrocyte sedimentation rate to 54 mm h^{-1}. Red cells and protein were now present in the urine. A skin biopsy from a purpuric area on his foot showed moderate edema (swelling) around the capillaries and in the dermis, as well as perivascular infiltrates of lymphocytes in the deeper dermis. Immunofluorescence microscopy of the biopsy tissue with the appropriate antibodies revealed the deposition of IgG and C3 in the perivascular areas.

Gregory was started on the anti-inflammatory corticosteroid prednisone, and all his symptoms improved progressively; the joint swelling and splenomegaly resolved over the next few days. He was soon able to walk and was discharged 7 days after the onset of his serum sickness on a slowly decreasing course of prednisone and Benadryl. On follow-up examination 2 weeks later, Gregory had no IgE antibodies against penicillin or ampicillin, as detected by both immediate hypersensitivity skin tests and by an *in vitro* fluorenzyme immunoassay. His parents were instructed that Gregory should never be given any penicillin, penicillin derivatives, or cephalosporins.

Route	Resulting disease	Site of immune complex deposition
Intravenous (high dose)	Vasculitis	Blood vessel walls
	Nephritis	Renal glomeruli
	Arthritis	Joint spaces
Subcutaneous	Arthus reaction	Perivascular area
Inhaled	Farmer's lung	Alveolar/capillary interface

Fig. 8.5 The dose and route of antigen delivery determine the pathology observed in type III allergic reactions.

Serum sickness.

The classic symptoms of serum sickness that Gregory showed were first described in great detail by Clemens von Pirquet and Bela Schick in a famous monograph entitled *Die Serumkrankheit* (serum sickness), published in 1905. Schick subsequently translated this monograph into English and it was reissued by Williams and Wilkins in 1951. It is fascinating to read this short work in the light of current knowledge. In the 1890s it had become common practice to treat diphtheria with horse serum containing antibodies taken from horses that had been immunized with diphtheria toxin. Immune horse serum was also used to treat scarlet fever, which was then a life-threatening illness. Von Pirquet and Schick made systematic observations on dozens of children who developed the symptoms and signs of serum sickness at the St Anna's Children's Hospital in Vienna and described the classic symptoms of the disease. They correctly surmised that serum sickness was due to an immunologic reaction to horse serum proteins in their patients (Fig. 8.7).

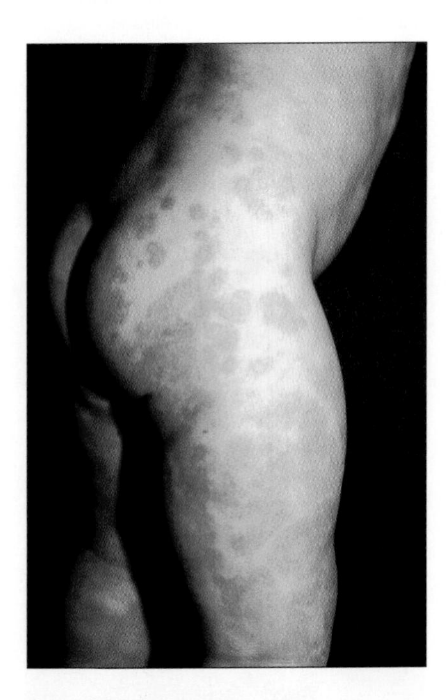

Fig.8.6 Urticarial rash as a consequence of a serum-sickness reaction.

Experimental models of serum sickness were developed in the 1950s by Hawn, Janeway, and Dixon, who injected rabbits with large amounts of bovine serum albumin or bovine gamma globulin. They noted that the rabbits developed glomerulonephritis just at the time when antibody against the foreign protein first appeared in the rabbit serum, accompanied by a profound and transitory fall in the serum complement level. By this time, immunochemistry had advanced to the point where it was possible to show that the disease was caused by the formation and deposition of small immune complexes.

Although horse serum is no longer used in therapy, other foreign proteins are still administered to patients. Antitoxins to snake venom are produced in various animal species, and mouse monoclonal antibodies are used in clinical practice. However, the commonest causes of serum sickness today are antibiotics, particularly penicillin and its derivatives, which act as haptens. These drugs bind to host proteins that serve as carriers and thus can elicit a rapid and strong IgG antibody response.

Serum sickness, although very unpleasant, is a self-limited disease that terminates as the immune response of the host moves into the zone of antibody excess. It can prove fatal if it provokes kidney shutdown or bleeding in a critical area such as the brain. Its course can be ameliorated by anti-inflammatory drugs such as prednisone and antihistamines. It is also unlike the other types of hypersensitivity in that a reaction can appear on first encounter with the antigen, if that is long-lived and given in a sufficiently large dose. This seems to have been the case for Gregory.

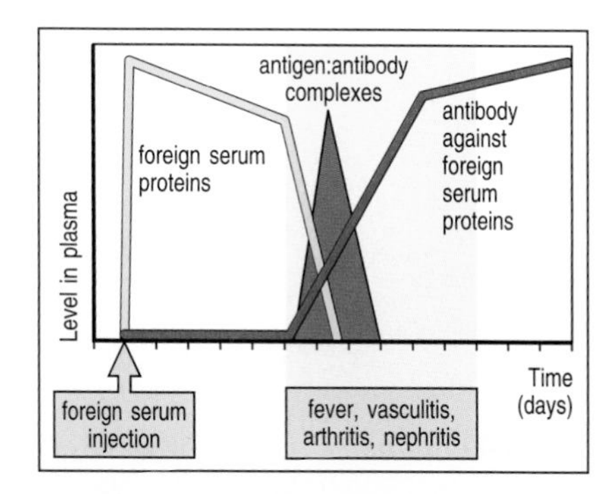

Fig. 8.7 Serum sickness is the classic example of a transient syndrome mediated by immune complexes. An injection of large amounts of foreign proteins, in this case derived from horse serum, leads to an antibody response. These antibodies form immune complexes with the circulating foreign proteins. These complexes activate complement and phagocytes, inducing fever, and are deposited in small blood vessels, inducing the symptoms of vasculitis, glomerulonephritis, and arthritis. All these effects are transient and resolve when the foreign protein is cleared from the system.

Questions.

1. *Hives (urticaria) and edema about the mouth and eyelids were the first symptoms of serum sickness developed by Gregory. What caused these early symptoms?*

2. *At one point Gregory became confused and disoriented and did not recognize his parents. His cerebrospinal fluid was normal and a CT scan of his brain was normal. However, an electroencephalogram displayed an abnormal pattern of brain waves. What produced these clinical and laboratory findings?*

3. *What other manifestations of vasculitis were noted in Gregory?*

4. *Gregory had enlarged lymph nodes everywhere and his spleen was also enlarged. If you had a biopsy of a node what would you expect to see?*

5. *Gregory had a brisk 'acute-phase response'. What is this, and what causes it?*

6. *Penicillin can cause more than one type of hypersensitivity reaction. What laboratory test gave the best evidence that Gregory was suffering from a disease caused by immune complexes?*

7. *When Gregory returned for a follow-up clinic visit, a skin test for immediate hypersensitivity was performed by intradermal injection of penicillin. He did not respond. Does this mean that an incorrect diagnosis was made and that he did not have serum sickness due to penicillin?*

CASE 9 Immune Dysregulation, Polyendocrinopathy, Enteropathy X-linked Disease

A failure of peripheral tolerance due to defective regulatory T cells.

The primary role of the immune system is to recognize pathogens and eliminate them from the body. However, an equally important task is to distinguish potentially dangerous antigens from those that are harmless. Countless innocuous foreign antigens are encountered every day by the lungs, gut, and skin, the interfaces between the body and the environment. Similarly, the body contains numerous self antigens that might bind to the specific antigen receptors on B and T cells. Activation of the immune system by such innocuous antigens is unnecessary and may lead to unwanted inflammation. Allergic and autoimmune diseases are well-known examples of such unwanted and potentially destructive responses.

Fortunately, unwanted immune responses are normally prevented or regulated by the phenomenon of immunologic tolerance. This is defined as nonresponsiveness of the lymphocyte population to the specific antigen, and arises at two stages of lymphocyte development. Central tolerance is the result of the removal of self-reactive lymphocytes in the central organs. Peripheral tolerance, in contrast, inactivates those T and B cells that escape central tolerance and exit to the periphery. Defects in either central or peripheral tolerance can result in unwanted or excessive immune responses.

Several mechanisms of peripheral tolerance exist (Fig. 9.1). One whose importance is increasingly being recognized is the network of regulatory cells that prevent or limit the activation of T cells, including self-reactive T cells, and the consequent destructive inflammatory processes. When these regulatory

Topics bearing on this case:
Peripheral tolerance
Regulatory T cells
Central tolerance
Autoimmunity

This case was prepared by Raif Geha, MD, in collaboration with Itai Pessach, MD.

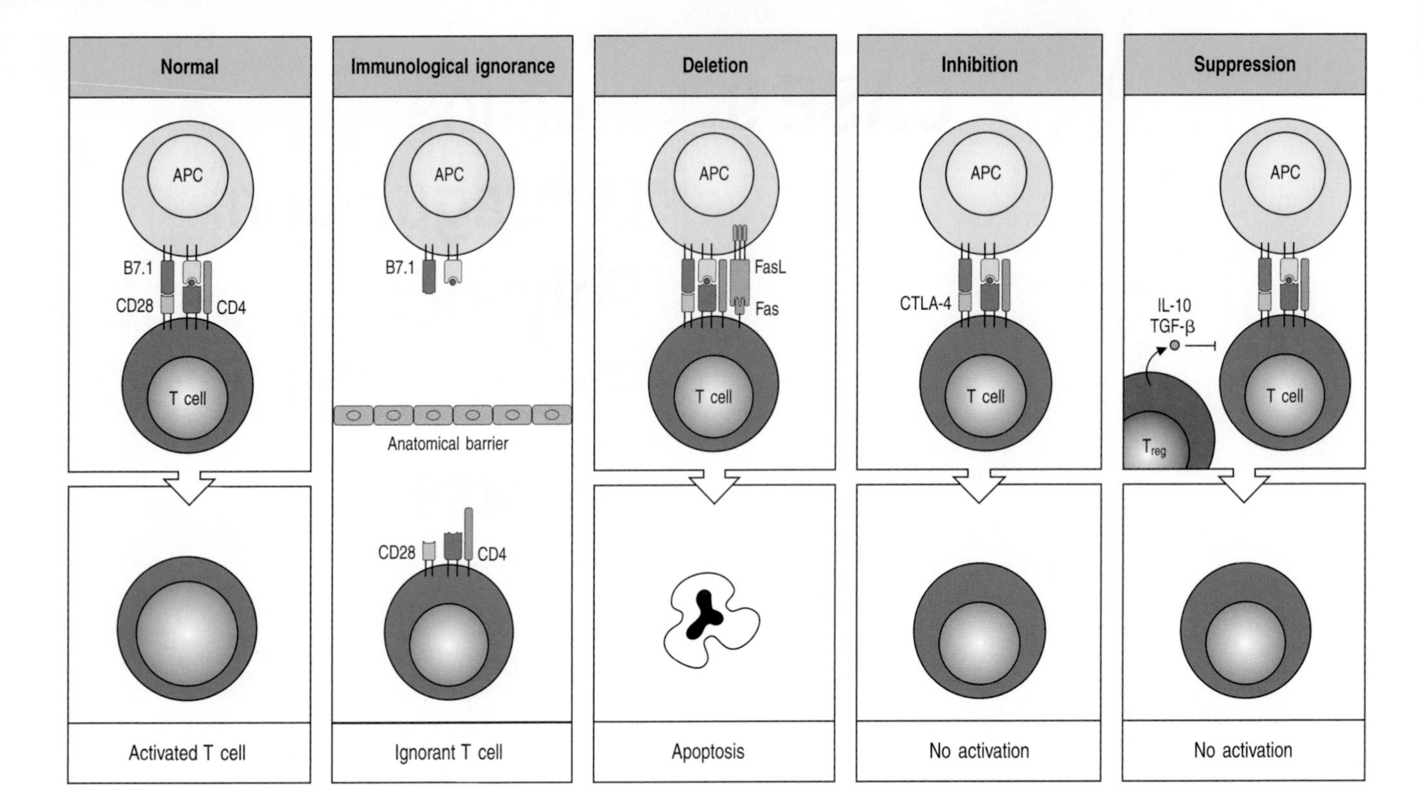

Fig. 9.1 Mechanisms of peripheral immunologic tolerance. T cells that are physically separated from their specific antigen—for example, by the blood–brain barrier—cannot become activated, a circumstance referred to as immunologic ignorance. T cells that express Fas (CD95) on their surface can receive signals from cells that express Fas ligand, leading to their deletion. The activation of naive T cells can be inhibited if the cell-surface protein CTLA-4 (CD152) binds B7.1 (CD80) on antigen-presenting cells (APCs). Regulatory T cells (mainly CD4 CD25 Foxp3-expressing) can inhibit, or suppress, other T cells, most probably through the production of inhibitory cytokines such as IL-10 and TGF-β.

cells do not function properly, problems can arise. A key cell type responsible for the maintenance of peripheral tolerance is the CD4 CD25 regulatory T cell (T_{reg}), also known as the natural regulatory T cell, which seems to become committed to a regulatory fate while still in the thymus and represents a small subset of circulating T cells (5–10%).

Although these cells were first characterized by their cell-surface CD25 (the α chain of the IL-2 receptor), this protein also appears on other T cells after activation. Natural T_{reg} cells are better characterized by their expression of the transcription factor Foxp3, which is essential for their specification and function as regulatory cells. Over the past decade they have emerged as crucial to the maintenance of peripheral tolerance. Neonatal thymectomy in mice and thymic hypoplasia in humans (DiGeorge syndrome) result in impaired generation of natural T_{reg} cells and the development of organ-specific autoimmune disease. The generation of natural T_{reg} cells in the thymus requires interaction with self-peptide:MHC class II complexes on cortical epithelial cells.

A second group of regulatory T cells seems to be induced from naive CD4 T cells in the periphery. These cells are $CD4^+$ $CD25^-$ and are heterogeneous, including subsets known as T_H3, T_R1, and a $CD4^+$ $CD25^-$ $Foxp3^+$ subset. Recently, a novel and rare population of T_{reg} cells that are CD8-positive has been described. Whereas CD4 T_{reg} subsets have been extensively studied, less is known about CD8 T_{reg} cells, their subsets, and their modes of action. NK cells and NKT cells have also been shown to be able to regulate immune responses. As a group, regulatory cells represent just one mechanism in a complex system of immunologic tolerance, acting to prevent or rein in unwanted immune responses.

The following case illustrates how a breakdown in peripheral tolerance as a result of a defect in regulatory T cells leads to a constellation of allergic symptoms, gastrointestinal symptoms, and autoimmune disease in infancy.

The case of Billy Shepherd: a defect in peripheral tolerance leading to dermatitis, diarrhea, and diabetes.

Billy was born at full term and developed atopic dermatitis (see Case 5) shortly after birth. This was treated by skin hydration and by the local application of hydrocortisone and antihistamines to control itching, the treatment being only partly successful. At 4 months of age, Billy developed an intractable watery diarrhea. Although he had initially gained weight well, by now his weight had fallen below the third centile for his age. At 6 months old, Billy started to develop high blood glucose levels and glucose in the urine. He was diagnosed with type 1 diabetes (insulin-dependent diabetes mellitus) and was referred by his pediatrician to the endocrine clinic at the Children's Hospital.

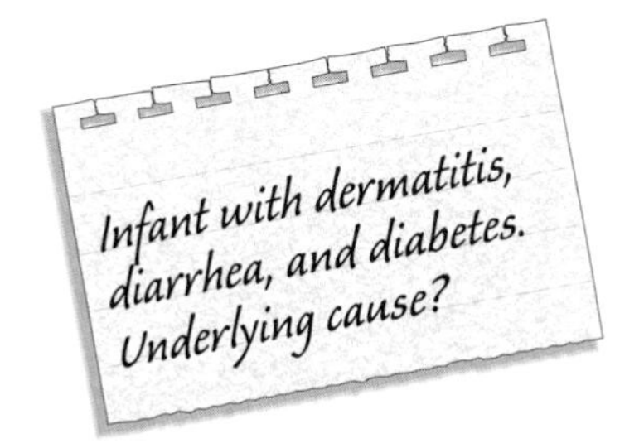

Infant with dermatitis, diarrhea, and diabetes. Underlying cause?

When first seen at the clinic, Billy weighed 5 kg (the third centile for age is 6.3 kg). He had diffuse eczema and sparse hair (Fig. 9.2). His cervical and axillary lymph nodes and spleen were enlarged. Laboratory tests revealed a normal white blood cell count of 7300 μl^{-1}, a normal hemoglobin of 11.3 g dl^{-1}, and a normal platelet count of 435,000 μl^{-1}. The percentage of eosinophils in the blood was high at 15% (normal <5%), and IgE was also elevated, at 1345 IU ml^{-1} (normal <50 IU ml^{-1}). Autoantibodies were found against glutamic acid decarboxylase (the GAD65 antigen) and against pancreatic islet cells. Billy was started on insulin therapy, which controlled his blood glucose level.

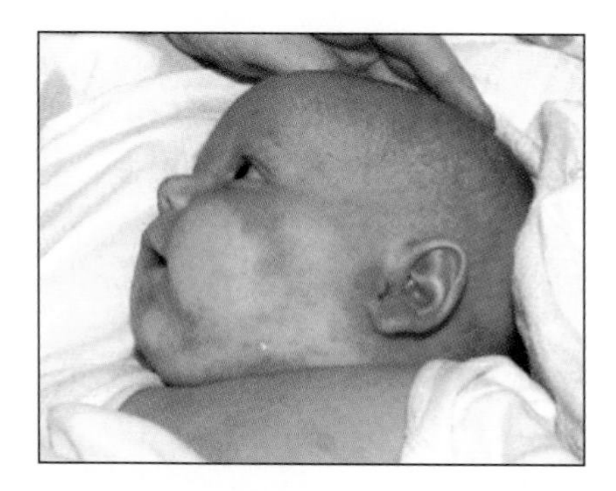

Fig. 9.2 Eczematous rash on the face of a baby boy with IPEX. Photograph courtesy of Talal Chatila, UCLA.

Because of the persistent diarrhea and failure to thrive, Billy required parenteral (intravenous) nutrition to maintain his weight. An endoscopy was ordered, to ascertain the cause of his persistent diarrhea, and a duodenal biopsy revealed almost total villous atrophy—an absence of villi in the lining of the duodenum—with a dense infiltrate of plasma cells and T cells (Fig. 9.3).

When Billy's mother was questioned, she revealed that there had been another son, who had died in infancy with severe diarrhea and a low platelet count. On the basis of Billy's symptoms and the family history, IPEX (immune dysregulation, polyendocrinopathy, enteropathy X-linked) was suspected. A FACS analysis of Billy's peripheral blood mononuclear cells revealed a lack of both CD4 CD25 cells and CD4 Foxp3-positive cells. Sequencing of Billy's *FOXP3* gene revealed a missense mutation, confirming the diagnosis.

With the diagnosis established, immunosuppressive therapy, including cyclosporin and tacrolimus, was started. Billy's diarrhea, glucose control, and eczema all improved markedly. After several months, however, his symptoms began to return and he stopped gaining weight. Shortly afterwards, he developed thrombocytopenia (a deficiency of blood platelets) and anti-platelet antibodies were detected.

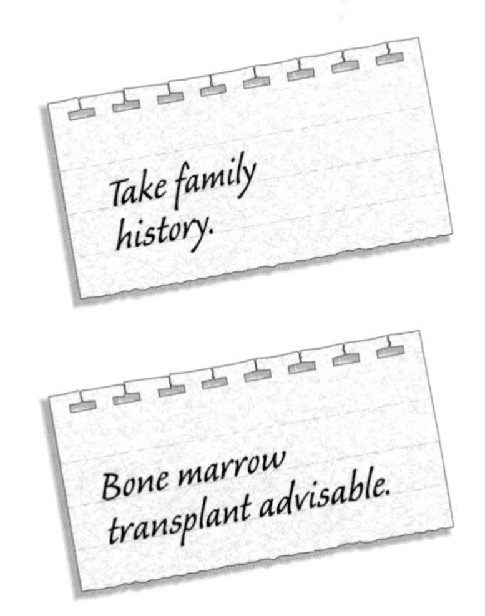

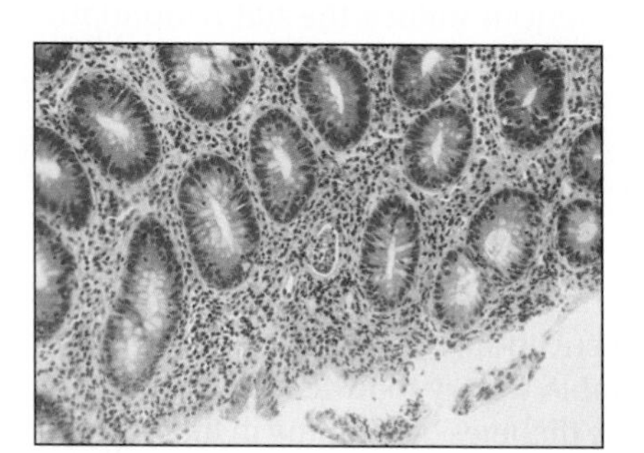

Fig. 9.3 Photomicrograph of duodenal biopsy from a child with IPEX. The section was stained with hematoxylin and eosin. Note the dense mononuclear cellular infiltrate. Photograph courtesy of Talal Chatila, UCLA.

The decision was made for Billy to be given a bone marrow transplant from his 5-year-old HLA-identical sister. In the weeks of conditioning leading up to the transplant, the diarrhea and eczema resolved. After transplantation, full engraftment of his sister's stem cells was established. Two weeks after transplantation, anti-platelet, anti-GAD65, and anti-islet cell antibodies could not be detected. A year after the transplant, Billy continues to be symptom-free, although analysis for chimerism reveals that only 30% of his T cells are derived from his sister's cells.

Immune dysregulation, polyendocrinopathy, enteropathy X-linked disease (IPEX).

IPEX is a very rare disease caused by mutations in the gene for the forkhead transcription factor Foxp3, which is essential for the function of CD4 CD25 T_{reg} cells. Foxp3 expression is restricted to a small subset of TCRα:β T cells and defines two pools of regulatory T cells: $CD4^+$ $CD25^{high}$ T cells and a minor population of $CD4^+$ $CD25^{lo/neg}$ T cells. Ectopic expression of *Foxp3 in vitro* and *in vivo* is sufficient to convert naive murine CD4 T cells to T_{reg} cells. In contrast, overexpression of *FOXP3* in naive human $CD4^+$ $CD25^-$ T cells *in vitro* will not generate potent suppressor activity, suggesting that additional factors are required. Foxp3 expression and suppressor function can, however, be induced in human $CD4^+$ $CD25^-$ $Foxp3^-$ cells by cross-linking of the T-cell receptor and stimulation via the co-stimulatory receptor CD28, or after antigen-specific stimulation. This suggests that *de novo* generation of T_{reg} cells in the periphery may be a natural consequence of the human immune response.

T_{reg} cells are anergic *in vitro*. They fail to secrete IL-2 or proliferate in response to ligation of their T-cell receptors, and depend on the IL-2 generated by activated CD4 T cells to survive and exert their function. An *in vitro* assay that measures the ability of CD4 CD25 T cells to suppress CD4 T-cell proliferation is commonly used to test for T_{reg} function (Fig. 9.4). How T_{reg} cells suppress immune responses *in vivo* is still unclear. There is some evidence for contact-dependent inhibition, whereas other studies suggest that regulatory T cells exert their function by secreting immunosuppressive cytokines such as IL-10 or transforming growth factor-β (TGF-β), or by directly killing their target cells in a perforin-dependent manner.

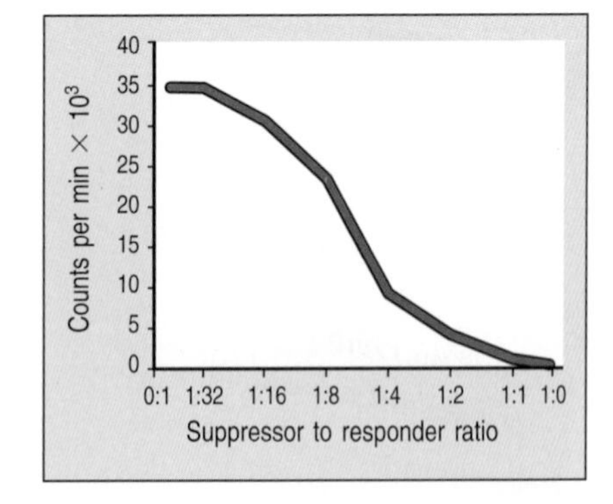

Fig. 9.4 Results of a functional assay for regulatory T cells from a normal individual. CD4 responder cells and CD4 CD25 T regulatory (suppressor) cells were mixed, together with antigen-presenting cells, at the ratios shown on the horizontal axis. The cells were stimulated with immobilized plate-bound anti-CD3 and soluble anti-CD28 for 3 days, then assessed for proliferation as measured by the incorporation of ^{3}H-labeled thymidine into DNA.

Several lines of evidence show that Foxp3 is crucial for the development and function of CD4 CD25 T_{reg} cells in mice. A mutation in *Foxp3* is responsible for an X-linked recessive inflammatory disease in the *Scurfy* mutant mouse. Male mice hemizygous for the mutation succumb to a CD4 T-cell-mediated lymphoproliferative disease characterized by wasting and multi-organ lymphocytic infiltration. T_{reg} cells are absent in *Scurfy* mice and in mice that have another spontaneous mutation in the *Foxp3* gene. In addition, specific ablation of *Foxp3* in T cells only is sufficient to induce the full lymphoproliferative autoimmune syndrome observed in the *Foxp3*-knockout mice. The Scurfy phenotype can be rescued by the introduction of a *Foxp3* transgene or by bone-marrow reconstitution, demonstrating the causative role of *Foxp3* in pathogenesis. Thus, the lack of *Foxp3*-expressing T_{reg} cells alone is sufficient to break self-tolerance and induce autoimmune disease.

In humans, missense or frameshift mutations in *FOXP3* result in loss of function of T_{reg} cells and uninhibited T-cell activation. As seen in Billy's case, the most common symptoms are an intractable watery diarrhea, leading to failure to thrive, dermatitis, and autoimmune diabetes developing in infancy. The diarrhea is due to widespread inflammation of the gut, including the colon (colitis), that results in villous atrophy, which reduces the absorptive capacity of the intestinal lining and thus contributes to wasting. Other diseases of immune dysregulation that are also seen include autoimmune thrombocytopenia, neutropenia, anemia, hepatitis, nephritis, hyperthyroidism or hypothyroidism, and food allergies. Autoantibodies also accompany these autoimmune diseases. Affected patients may also suffer more frequent infections, including sepsis, meningitis, or pneumonia, although the reason for the increased susceptibility to infection is unclear. Patients generally have normal immunoglobulin levels (except for the elevated IgE), and their ability to make specific antibody is intact.

Questions.

1 *Billy continued to remain asymptomatic even though he eventually only had around 30% engraftment of his sister's bone marrow stem cells (as judged from the proportions of T cells). Why is full engraftment not necessary in patients with IPEX?*

2 *Why did Billy's diarrhea improve while he was being prepared for transplantation?*

3 *Intravenous immunoglobulin (IVIG) has been used to treat IPEX. How might this be an effective therapy?*

4 *The occurrence of colitis in IPEX suggests that T_{reg} cells may be implicated in its pathogenesis and that they might be used therapeutically in the more common forms of colitis. Is there experimental data to support this claim?*

5 *What other gene mutations can give rise to a clinical picture similar to IPEX?*

CASE 10 Hereditary Angioedema

Regulation of complement activation.

Complement is a system of plasma proteins that participates in a cascade of reactions, generating active components that allow pathogens and immune complexes to be destroyed and eliminated from the body. Complement is part of the innate immune defenses of the body and is also activated via the antibodies produced in an adaptive immune response. Complement activation is generally confined to the surface of pathogens or circulating complexes of antibody bound to antigen.

Complement is normally activated by one of three routes: the classical pathway, which is triggered by antigen:antibody complexes or antibody bound to the surface of a pathogen; the lectin pathway, which is activated by mannose-binding lectin (MBL) and the ficolins; and the alternative pathway, in which complement is activated spontaneously on the surface of some bacteria. The early part of each pathway is a series of proteolytic cleavage events leading to the generation of a convertase, a serine protease that cleaves complement component C3 and thereby initiates the effector actions of complement. The C3 convertases generated by the three pathways are different, but evolutionarily homologous, enzymes. Complement components and activation pathways, and the main effector actions of complement, are summarized in Fig. 10.1.

The principal effector molecule, and a focal point of activation for the system, is C3b, the large cleavage fragment of C3. If active C3b, or the homologous but less potent C4b, accidentally becomes bound to a host cell surface instead of a pathogen, the cell can be destroyed. This is usually prevented by the rapid hydrolysis of active C3b and C4b if they do not bind immediately to the surface where they were generated. Protection against inappropriate activation of complement is also provided by regulatory proteins.

One of these, and the most potent inhibitor of the classical pathway, is the C1 inhibitor (C1INH). This belongs to a family of serine protease inhibitors (called serpins) that together constitute 20% of all plasma proteins. In addition to being the sole known inhibitor of C1, C1INH contributes to the

Topics bearing on this case:
Classical pathway of complement activation
Inhibition of C1 activation
Alternative pathway of complement activation
Inflammatory effects of complement activation
Regulation of C4b

This case was prepared by Raif Geha, MD, in collaboration with Arturo Borzutzky, MD.

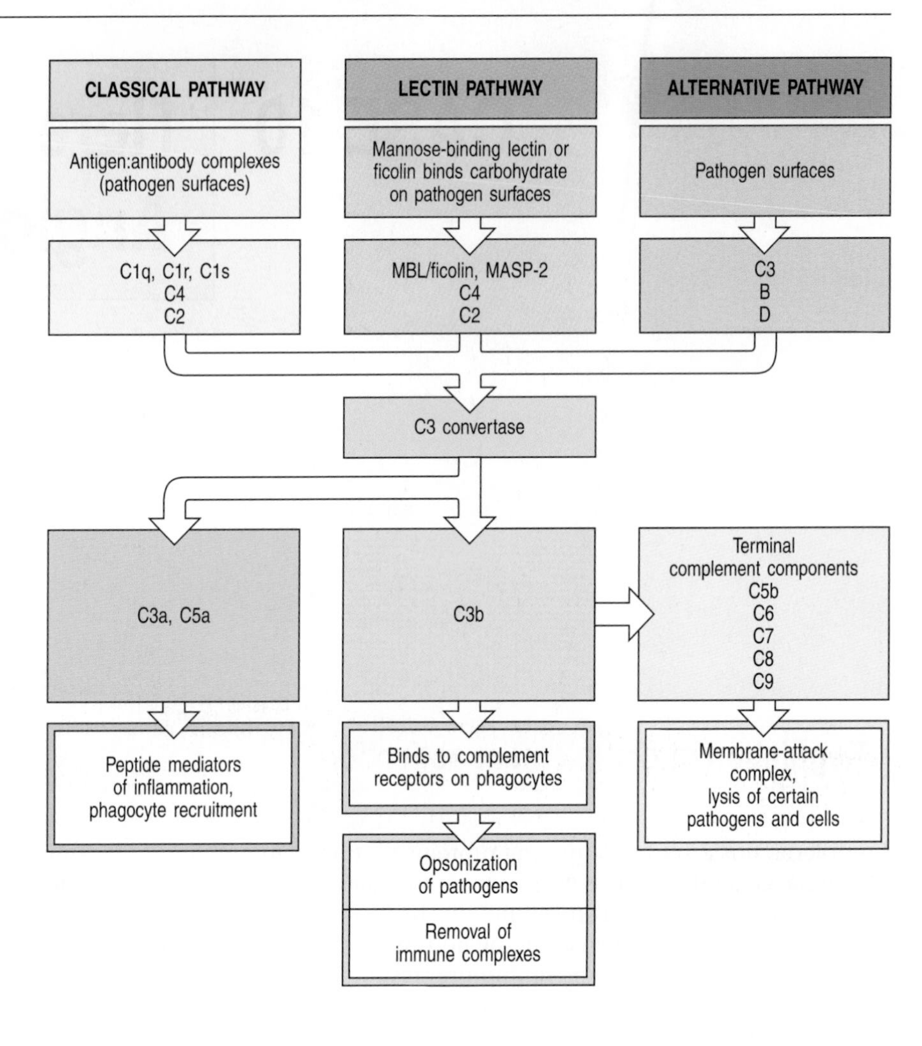

Fig. 10.1 Overview of the main components and effector actions of complement. The early events of all three pathways of complement activation involve a series of cleavage reactions that culminate in the formation of an enzymatic activity called a C3 convertase, which cleaves complement component C3 into C3b and C3a. The production of the C3 convertase is the point at which the three pathways converge and the main effector functions of complement are generated. C3b binds covalently to the bacterial cell membrane and opsonizes the bacteria, enabling phagocytes to internalize them. C3a is a peptide mediator of local inflammation. C5a and C5b are generated by the cleavage of C5b by a C5 convertase formed by C3b bound to the C3 convertase (not shown in this simplified diagram). C5a is also a powerful peptide mediator of inflammation. C5b triggers the late events in which the terminal components of complement assemble into a membrane-attack complex that can damage the membrane of certain pathogens. Although the classical complement activation pathway was first discovered as an antibody-triggered pathway, it is now known that C1q can activate this pathway by binding directly to pathogen surfaces, as well as paralleling the lectin activation pathway by binding to antibody that is itself bound to the pathogen surface. In the lectin pathway, MASP stands for mannose-binding lectin-associated serine protease.

regulation of serine proteases of the clotting system and of the kinin system, which is activated by injury to blood vessels and by some bacterial toxins. The main product of the kinin system is bradykinin, which causes vasodilation and increased capillary permeability.

C1INH intervenes in the first step of the complement pathway, when C1 binds to immunoglobulin molecules on the surface of a pathogen or antigen:antibody complex (Fig. 10.2). Binding of two or more of the six tulip-like heads of the C1q component of C1 is required to trigger the sequential activation of the two associated serine proteases, C1r and C1s. C1INH inhibits both of these proteases, by presenting them with a so-called bait-site, in the form of an arginine bond that they cleave. When C1r and C1s attack the bait-site they covalently bind C1INH and dissociate from C1q. By this mechanism, the C1 inhibitor limits the time during which antibody-bound C1 can cleave C4 and C2 to generate C4b2a, the classical pathway C3 convertase.

Activation of C1 also occurs spontaneously at low levels without binding to an antigen:antibody complex, and can be triggered further by plasmin, a protease of the clotting system, which is also normally inhibited by C1INH. In the absence of C1INH, active components of complement and bradykinin are produced. This is seen in hereditary angioedema (HAE), a disease caused by a genetic deficiency of C1INH.

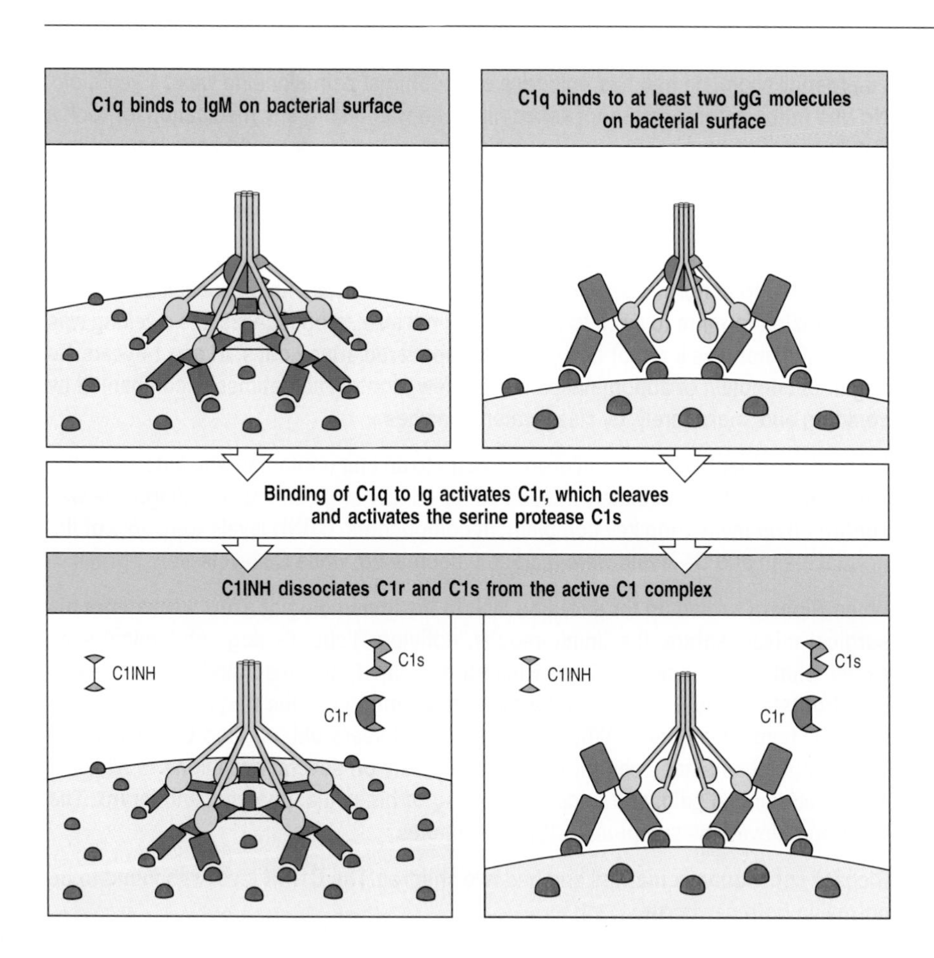

Fig. 10.2 Activation of the classical pathway of complement and intervention by C1INH. In the left panel, one molecule of IgM, bent into the 'staple' conformation by binding several identical epitopes on a pathogen surface, allows binding by the globular heads of C1q to its Fc pieces on the surface of the pathogen. In the right panel, multiple molecules of IgG bound to the surface of the pathogen allow binding by C1q to two or more Fc pieces. In both cases, binding of C1q activates the associated C1r, which becomes an active enzyme that cleaves the proenzyme C1s, a serine protease that initiates the classical complement cascade. Active C1 is inactivated by C1INH, which binds covalently to C1r and C1s, causing them to dissociate from the complex. There are in fact two C1r and two C1s molecules bound to each C1q molecule, although for simplicity this is not shown here. It takes four molecules of C1INH to inactivate all the C1r and C1s.

The case of Richard Crafton: a failure of communication as well as of complement regulation.

Richard Crafton was a 17-year-old high-school senior when he had an attack of severe abdominal pain at the end of a school day. The pain came as frequent sharp spasms and he began to vomit. After 3 hours, the pain became unbearable and he went to the emergency room at the local hospital.

Richard, age 17, presents as an acute abdominal emergency.
Appendectomy performed. Appendix appears normal.

At the hospital, the intern who examined him found no abnormalities other than dry mucous membranes of the mouth, and a tender abdomen. There was no point tenderness to indicate appendicitis. Richard continued to vomit every 5 minutes and said the pain was getting worse.

A surgeon was summoned. He agreed with the intern that Richard had an acute abdominal condition but was uncertain of the diagnosis. Blood tests showed an elevated red blood cell count, indicating dehydration. The surgeon decided to proceed with exploratory abdominal surgery. A large midline incision revealed a moderately swollen and pale jejunum but no other abnormalities were noted. The surgeon removed Richard's appendix, which was normal, and Richard recovered and returned to school 5 days later.

What Richard had not mentioned to the intern or to the surgeon was that, although he had never had such severe pains as those he was experiencing when he went to the

Family history
of colic.

emergency room, he had had episodes of abdominal pain since he was 14 years old. No one in the emergency room asked him if he was taking any medication, or took a family history or a history of prior illness. If they had, they would have learned that Richard's mother, his maternal grandmother, and a maternal uncle, also had recurrent episodes of severe abdominal pain, as did his only sibling, a 19-year-old sister.

As a newborn, Richard was prone to severe colic. When he was 4 years old, a bump on his head led to abnormal swelling. When he was 7, a blow with a baseball bat caused his entire left forearm to swell to twice its normal size. In both cases, the swelling was not painful, nor was it red or itchy, and it disappeared after 2 days. At age 14 years, he began to complain of abdominal pain every few months, sometimes accompanied by vomiting and, more rarely, by clear, watery diarrhea.

Richard's mother had taken him at age 4 years to an immunologist, who listened to the family history and immediately suspected hereditary angioedema. The diagnosis was confirmed on measuring key complement components. C1INH levels were 16% of the normal mean and C4 levels were markedly decreased, while C3 levels were normal.

When Richard turned up for a routine visit to his immunologist a few weeks after his surgical misadventure, the immunologist, noticing Richard's large abdominal scar, asked what had happened. When Richard explained, he prescribed daily doses of Winstrol (stanozolol). This caused a marked diminution in the frequency and severity of Richard's symptoms. When Richard was 20 years old, purified C1INH became available; he has since been infused intravenously on several occasions to alleviate severe abdominal pain, and once for swelling of his uvula, pharynx, and larynx. The infusion relieved his symptoms within 25 minutes.

Richard subsequently married and had two children. The C1INH level was found to be normal in both newborns.

Hereditary angioedema.

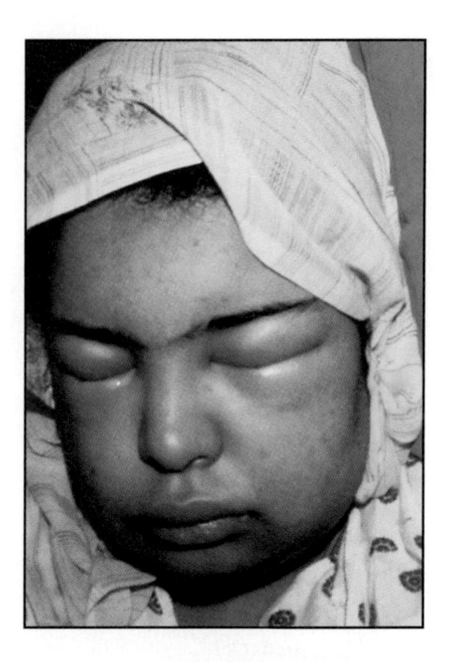

Fig. 10.3 Hereditary angioedema. Transient localized swelling that occurs in this condition often affects the face.

Individuals like Richard with a hereditary deficiency of C1INH are subject to recurrent episodes of circumscribed swelling of the skin (Fig. 10.3), intestine, and airway. Attacks of subcutaneous or mucosal swelling most commonly affect the extremities, but can also involve the face, trunk, genitals, lips, tongue, or larynx. Cutaneous attacks cause temporary disfigurement but are not dangerous. When the swelling occurs in the intestine it causes severe abdominal pain, and obstructs the intestine so that the patient vomits. When the colon is affected, watery diarrhea may occur. Swelling in the larynx is the most dangerous symptom, because the patient can rapidly choke to death. HAE attacks do not usually involve itching or hives, which is useful to differentiate this disease from allergic angioedema. However, a serpiginous, or linear and wavy, rash is sometimes seen before the onset of swelling symptoms. Such episodes may be triggered by trauma, menstrual periods, excessive exercise, exposure to extremes of temperature, mental stress, and some medications such as angiotensin-converting enzyme inhibitors and oral contraceptives.

HAE is not an allergic disease, and attacks are not mediated by histamine. HAE attacks are associated with activation of four serine proteases, which are normally inhibited by C1INH. At the top of this cascade is Factor XII, which directly or indirectly activates the other three (Fig. 10.4). Factor XII is normally activated by injury to blood vessels, and initiates the kinin cascade, activating

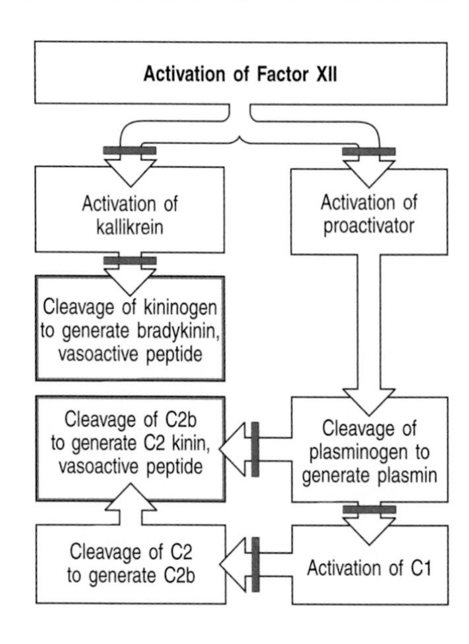

Fig. 10.4 Pathogenesis of hereditary angioedema. Activation of Factor XII leads to the activation of kallikrein, which cleaves kininogen to produce the vasoactive peptide bradykinin; it also leads to the activation of plasmin, which in turn activates C1. C1 cleaves C2, whose smaller fragment C2b is further cleaved by plasmin to generate the vasoactive peptide C2 kinin. The red bars represent inhibition by C1INH.

kallikrein, which generates the vasoactive peptide bradykinin. Factor XII also indirectly activates plasmin, which, as mentioned earlier, activates C1 itself. Plasmin also cleaves C2b to generate a vasoactive fragment called C2 kinin. In patients deficient in C1INH, the uninhibited activation of Factor XII leads to the activation of kallikrein and plasmin; kallikrein catalyzes the formation of bradykinin, and plasmin produces C2 kinin. Bradykinin is the main mediator responsible for HAE attacks by causing vasodilation and increasing the permeability of the postcapillary venules by causing contraction of endothelial cells so as to create gaps in the blood vessel wall (Fig. 10.5). This is responsible for the edema; movement of fluid from the vascular space into another body compartment, such as the gut, causes the symptoms of dehydration as the vascular volume contracts.

Treatment of HAE can focus on preventing attacks or on resolving acute episodes. Purified or recombinant C1INH is an effective therapy in both these settings. A kallikrein inhibitor and a bradykinin receptor antagonist have also been developed to target the kinin cascade and bradykinin activity.

Questions.

1 *Activation of the complement system results in the release of histamine and chemokines, which normally produce pain, heat, and itching. Why is the edema fluid in HAE free of cellular components, and why does the swelling not itch?*

2 *Richard has a markedly decreased amount of C4 in his blood. This is because it is being rapidly cleaved by activated C1. What other complement component would you expect to find decreased? Would you expect the alternative pathway components to be low, normal, or elevated? What about the terminal components?*

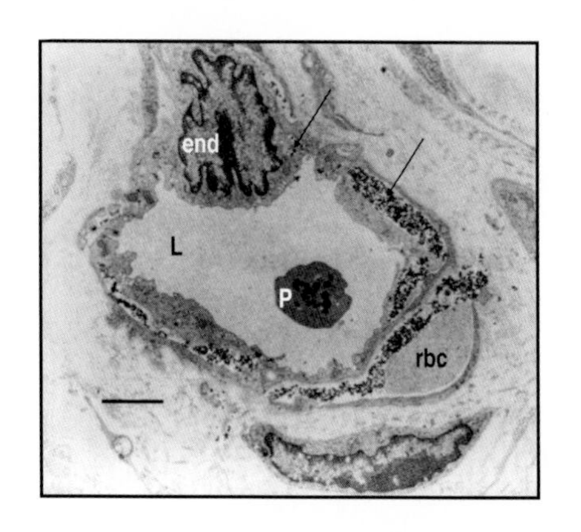

Fig. 10.5 Contraction of endothelial cells creates gaps in the blood vessel wall. A guinea pig was injected intravenously with India ink (a suspension of carbon particles). Immediately thereafter the guinea pig was injected intradermally with a small amount of activated C1s. An area of angioedema formed about the injected site, which was biopsied 10 minutes later. An electron micrograph reveals that the endothelial cells in post-capillary venules have contracted and formed gaps through which the India ink particles have leaked from the blood vessel. L is the lumen of the blood vessel; P is a polymorphonuclear leukocyte in the lumen; rbc is a red blood cell that has leaked out of the blood vessel. Micrograph courtesy of Kaethe Willms.

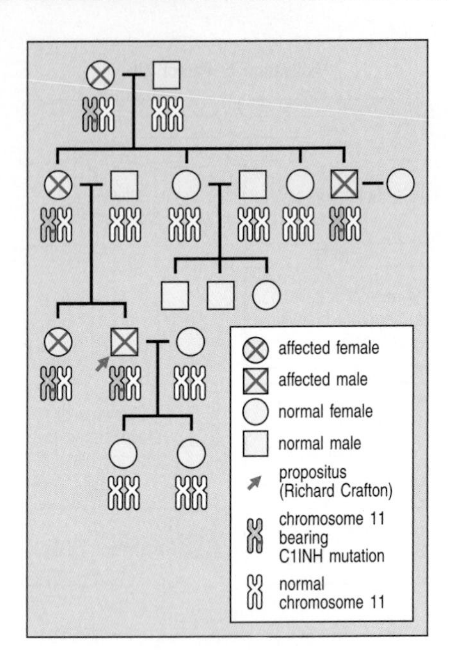

Fig. 10.6 The inheritance of hereditary angioedema in Richard's extended family.

3 *Despite the complement deficiency in patients with HAE, they are not unduly susceptible to infection. Why not?*

4 *What is stanozolol, and why was it prescribed?*

5 *Emergency treatment for HAE cases is sometimes necessary because of airway obstruction. In most cases, however, a patient with obstruction of the upper airways is likely to be suffering from an anaphylactic reaction. The treatment in this case would be epinephrine. How might you decide whether to administer epinephrine or intravenous C1INH?*

6 *Figure 10.6 shows Richard's family tree. What is the mode of inheritance (dominant or recessive, sex-linked or not) of HAE? Can Richard's two children pass the disease onto their offspring?*

CASE 11 Urticaria

An itchy rash caused by mast-cell activation in the skin.

Urticaria, a rash commonly known as hives, occurs when dermal mast cells are activated, resulting in the release of potent allergic mediators. These induce increased vascular permeability with extravasation of fluid and the formation of discrete areas of edema in the dermis, resulting in the formation of wheals. The mediators released by mast cells also trigger vasodilation, leading to enhanced blood flow, and give rise to localized areas of non-raised erythema (flare) in the classic wheal-and-flare response (Fig. 11.1). The dermal mast-cell activation that underlies urticaria is a common endpoint of several distinct pathophysiological processes.

Immune-mediated urticaria, the most common form, results from IgE-mediated mast-cell activation (Fig. 11.2). This occurs in individuals who have been previously sensitized to produce specific IgE antibodies against allergens including foods, insect venoms, and drugs. These allergen-specific IgE antibodies are tightly bound to dermal mast cells through the high-affinity IgE receptor, FcεRI. Upon allergen encounter, the receptor-bound IgE induces FcεRI aggregation, triggering a signaling cascade that causes the mast cells to degranulate and release preformed inflammatory mediators. These result in the clinical manifestations of urticaria. Resolution occurs within hours to days after cessation of exposure to the triggering antigen. In some circumstances, IgG antibodies can also induce mast-cell activation. Mast cells express the IgG receptor FcγRIII, which has a much lower affinity for IgG than FcεRI has for

Urticarial lesions arising from a skin test for allergy

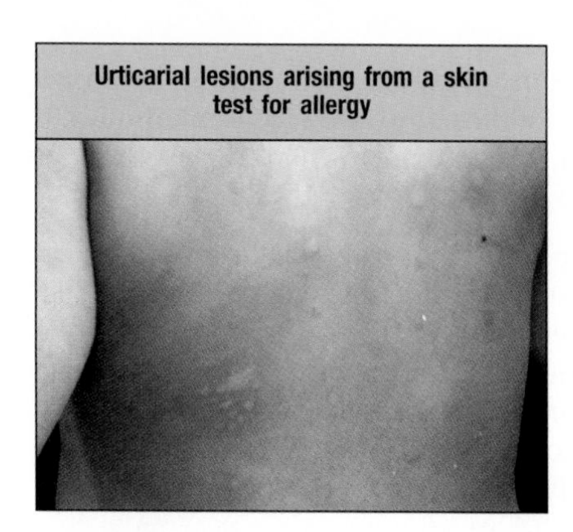

Fig. 11.1 Urticarial lesions arising from a skin test for allergy. Hives are heterogeneous in size and shape and have a characteristic wheal (raised center) due to edema, and erythema (redness) due to vasodilation. The centers can have a blanched appearance.

Topics bearing on this case:

- Mast cells
- Mast-cell activation
- Mast-cell mediators

This case was prepared by Hans Oettgen, MD, PhD, and Raif Geha, MD, in collaboration with Christina Yee, MD, PhD.

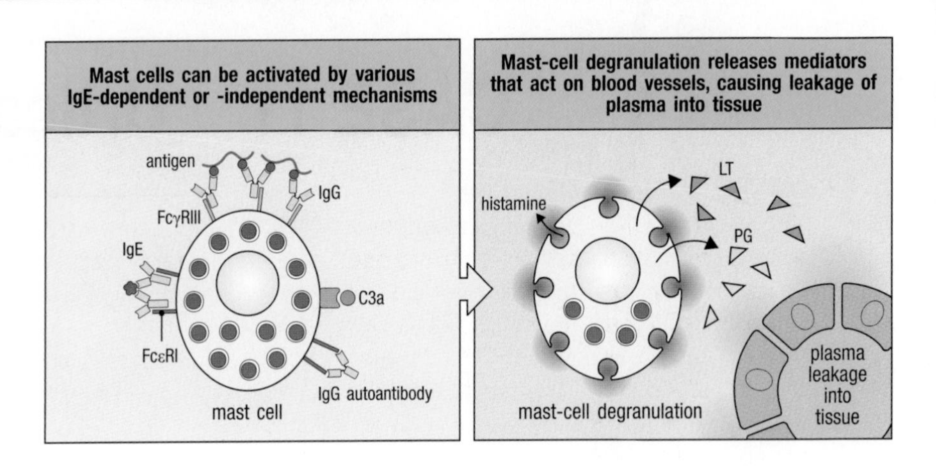

Fig. 11.2 Pathways of mast-cell activation in urticaria caused by immunological mechanisms. Dermal mast cells bearing allergen-specific IgE antibodies bound to the high-affinity IgE receptor, FcεRI, can be activated after allergen encounter and resultant receptor cross-linking (left panel). Alternatively, IgG antibodies can bind to the IgG receptor FcγRIII on mast cells. This is a relatively low-affinity interaction that requires clustering of IgG antibodies in immune complexes with antigen. Immune complexes can also activate the classical complement pathway, leading to the generation of the complement anaphylatoxins (C3a and C5a), which can activate mast cells via their respective receptors. In some patients, the production of autoantibodies that bind to FcεRI provides a trigger. Activated mast cells release preformed vasoactive mediators, which include histamine and the lipid metabolites of arachidonic acid—leukotrienes (LT) and prostaglandins (PG) (right panel). These mediators induce local vasodilation, with resultant increased blood flow and erythema, and increased endothelial permeability, leading to plasma extravasation and edema.

IgE. As a result, whereas FcεRI is constitutively saturated with monomeric IgE at physiologic IgE levels, FcγRIII binding to IgG is favored only when the IgG is in polyvalent form, as would occur in immune complexes, which can arise in settings of relatively high and roughly equivalent concentrations of antibody and antigen. In the clinical syndrome of serum sickness (Case 8), first described in patients receiving horse immune serum as a therapeutic agent but now recognized as an infrequent complication of drug therapy and infections, IgG antibodies produced in high-titers 2–3 weeks after exposure interact with antigen to form immune complexes. These can efficiently activate mast cells both through engagement of FcγRIII and through complement fixation.

Complement-mediated urticaria occurs when activated complement components bind to mast cells. Circulating immune complexes activate the complement cascade to produce C3a and C5a fragments that engage specific receptors on mast cells. Serum sickness (see Case 8) and the autoimmune disease systemic lupus erythematosus can be associated with the production of immune complexes that trigger the urticaria. Other causes of increased complement activation that result in urticaria include acquired angioedema, arising from the acquired deficiency of C1 inhibitor (commonly associated with lymphoproliferative disorders and other malignancies), and vasculitis.

Non-immune-mediated urticaria can be caused by the direct activation of dermal mast cells. Physical stimuli, including cold, pressure, heat, and vibration, can trigger mast-cell activation in susceptible individuals. Certain classes of medications can also activate mast cells directly, most notably opiates (such as morphine or codeine) and the antibiotic vancomycin, as can some foods (strawberries, tomatoes, and shellfish) and chemicals (radiocontrast dyes and ethanol).

Autoimmune urticaria occurs in individuals who have circulating autoantibodies specific for FcεRI. These autoantibodies bind and cross-link the FcεRI receptor, leading to mast-cell activation. It is not known what triggers the production of these autoantibodies.

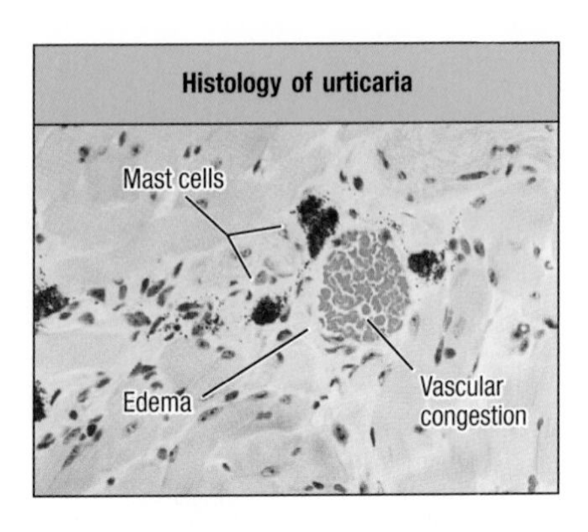

Fig. 11.3 Histology of urticaria. Degranulating mast cells are seen in this section with numerous granules visible outside the confines of the cell membrane. Vascular congestion and edema (white intracellular spaces) are also evident.

Microscopic examination of skin biopsies from urticarial lesions demonstrates edema and dilatation of the small blood vessels in the dermis (Fig. 11.3). Perivascular lymphocytes, swollen collagen fibers, and flattened rete pegs (epithelial extensions into the underlying connective tissue) may also be observed within the dermis. In chronic urticaria, when lesions persist for weeks or months, activated inflammatory cells are recruited to the affected skin, and a dense dermal infiltrate of lymphocytes, neutrophils, and eosinophils can be observed.

The case of Fitzwilliam Darcy: a mysterious rash after a visit to the grandparents.

Fitzwilliam Darcy, a 7-year-old boy, was brought to the Children's Hospital primary-care clinic by his parents because of a sudden-onset rash. They became alarmed when he woke up from a nap in the afternoon with a bright red rash over his arms and legs. The rash consisted of raised blotchy lesions, some with the appearance of welts. He was scratching intensely and appeared very uncomfortable. The parents noticed that some of the red spots would disappear spontaneously, with new spots arising in different locations. The rash seemed to be rapidly evolving. His parents also noticed that Fitzwilliam's eyelids were swollen, but that he did not have any difficulty breathing. The child had spent the morning with his grandparents at the Science Museum and playing outside in their yard.

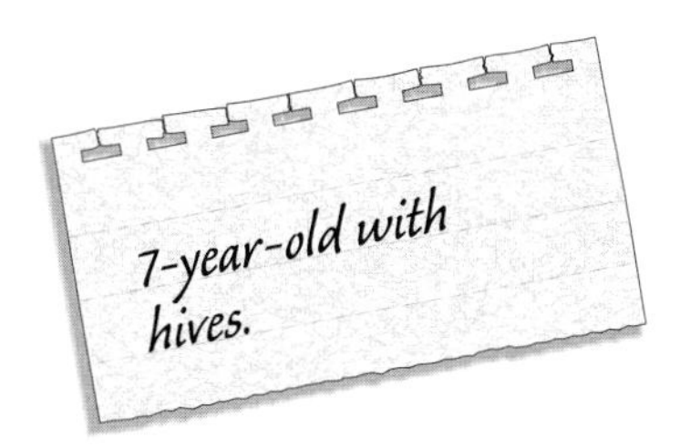

Fitzwilliam had a history of food allergy to peanuts. His parents remembered that he had had a similar rash when he was 2 years old. He had eaten a peanut butter sandwich and developed wheezing, lip swelling, facial swelling, and a few hives. His pediatrician ordered a test for IgE antibodies against peanut extract; the child was found to have peanut-specific IgE antibodies, and a diagnosis of peanut allergy was made. Ever since, his family had been extremely careful about keeping him away from peanuts, and they did not think that eating peanuts was the cause of his current rash.

A review of the family history revealed that Fitzwilliam also had an aunt who suffered from frequent outbreaks of hives. She had been taking a medication to prevent the hives and they would recur whenever she stopped taking it. Her evaluation had not shown any evidence of allergies but revealed abnormalities in thyroid function.

The pediatrician observed that Fitzwilliam appeared well, but itchy. He had about 10 erythematous, raised red plaques, a few millimeters to several centimeters in size, of varying shapes, and scattered on the exposed areas of his arms and legs. Some of these hives had a ring-like appearance and many had a surrounding flare of erythema. The lesions were swollen, but did not feel indurated (hard). Excoriations (scratch marks) were evident around some of the hives. He had one on his right cheek, and his right eye was a little swollen, but his lips and tongue were not swollen. His lungs were clear and his heart rate was regular.

Fitzwilliam was given a dose of the oral antihistamine diphenhydramine, and within an hour his hives started to fade. He remained well until 3 days later, when he went to his grandparents' house again and developed hives on his arms and face, this time even before he had returned home. While he was there, his grandparents found him playing with their neighbor's cat, and Fitzwilliam admitted he had secretly petted the cat on the day he first developed hives. He was again treated with diphenhydramine with good effect.

Fitzwilliam's pediatrician referred him to the Allergy Clinic at Children's Hospital for evaluation of possible cat allergy. Skin-prick allergy testing revealed positive reactions to cat dander and grass pollen. Blood testing revealed high IgE levels along with significant levels of IgE antibodies specific for cat dander, grass pollen, and peanuts. Fitzwilliam's parents were advised to keep him away from cats and to limit his exposure to grass pollens. On the allergist's advice, his family started giving him a dose of antihistamine before anticipated contact with cats. He had no further episodes of hives even after he played with the neighbor's cat at his grandparents during a visit a month later.

Positive skin tests for allergy to cat dander.

Urticaria.

It has been estimated that one in five individuals in the United States will experience at least one episode of urticaria in their lifetime. However, identifying the root cause of the urticaria can pose significant challenges. It is a source of frustration for patients and health-care providers that in many cases the underlying trigger is never identified.

Urticaria presents as a rash with pruritic (itchy), erythematous, raised edematous lesions, commonly called hives. Urticarial lesions have a very characteristic appearance—typically well circumscribed, blanching on pressure—and appear anywhere on the body, especially at sites of allergen exposure or pressure on the skin. Classically, the response is described as a wheal (raised lesion) with surrounding flare (flat erythema). As observed in Fitzwilliam, the rash is constantly changing, with new lesions appearing and old lesions resolving.

The initial classification of urticaria is based on duration. Acute urticaria is defined as the presence of urticaria for less than 6 weeks, whereas chronic urticaria lasts 6 weeks or longer.

Acute urticaria is most often immunologically mediated, with the most commonly identified triggers being foods, medications, insect venoms, and environmental allergens. Local urticarial reactions can be triggered by direct allergen contact with the skin. Urticaria precipitated by allergen ingestion, inhalation, or injection can be restricted to the skin, or be part of a systemic hypersensitivity reaction (anaphylaxis) involving multiple organ systems (see Case 1). In children, viral infections are a common cause of acute urticaria, but often the specific immunological trigger is unknown. Acute urticaria is generally self-limited, sometimes lasting just an hour or two. Most cases are treated by removing the trigger (if known) and alleviating the symptoms until the urticaria resolves. Laboratory evaluation is not indicated unless food allergy is suspected. Because the severity of reactions to foods can increase with recurrent exposure and evolve to systemic anaphylaxis, it is important to identify food sensitivity so that guidance can be given on avoidance of allergen and emergency treatment of reactions.

Chronic urticaria occurs more often in adults than in children. Allergic sensitivity to foods, medications, and environmental allergens should be investigated, but allergic triggers are identified in less than 20% of cases of urticaria lasting more than 6 weeks. An estimated 10–20% of chronic urticaria cases can be classified as physical urticaria, in which exposure to extreme temperatures (cold or heat), pressure, or vibration is the trigger for dermal mast-cell activation. Autoimmune diseases, most commonly thyroid disease, can also be associated with chronic urticaria. In cases of chronic urticaria, laboratory evaluation is commonly performed for antibodies against thyroid antigens and FcεRI as well as for the anti-nuclear and anti-double-stranded DNA antibodies characteristic of systemic lupus erythematosus (Fig. 11.4).

The occurrence of urticaria in individuals with autoimmunity may be related to several factors. A predilection to produce anti-FcεRI antibodies, the presence of circulating immune complexes (which can activate mast cells both by means of FcγRIII and by generation of the complement anaphylatoxins C3a and C5a), and inflammation (vasculitis) of small blood vessels in the skin have all been hypothesized to contribute to the association of urticaria with autoimmunity. Despite extensive evaluation, the etiology of chronic urticaria is established in only a minority of cases. Fortunately, chronic urticaria commonly resolves spontaneously over weeks to months, with most patients experiencing gradual remission of their symptoms within 3–5 years.

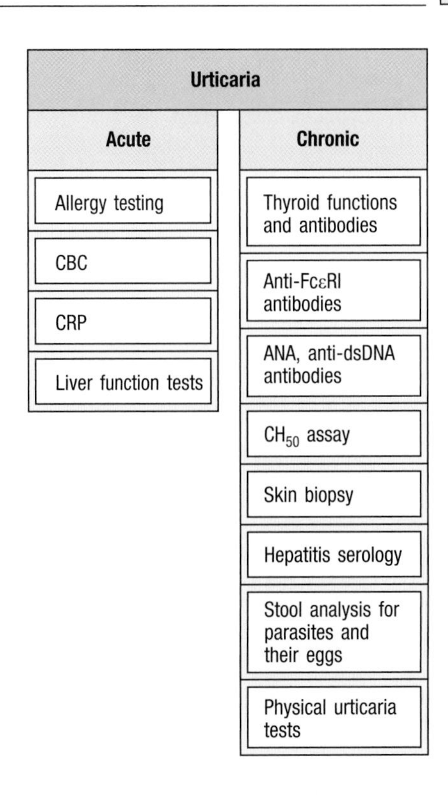

Fig. 11.4 Evaluation of urticaria. For acute urticaria (hives of less than 6 weeks duration), testing for IgE antibodies specific for food or environmental allergens can be very helpful. A complete blood count (CBC) with differential diagnosis can provide clues to infectious or allergic triggers. For example, lymphocytosis and/or the presence of atypical lymphocytes can occur with Epstein–Barr virus infection (infectious mononucleosis), whereas an elevated eosinophil count might suggest allergic triggers or parasite infestation. C-reactive protein (CRP) is a useful screen for inflammatory processes, including vasculitis, and liver function tests can detect viral hepatitis, which can present with urticaria as a result of the generation of immune complexes containing viral antigens. For chronic urticaria (more than 6 weeks duration), assessment for anti-thyroid autoantibodies is recommended, because the presence of these antibodies in thyroiditis can be accompanied by hives. Some patients produce IgG antibodies against FcεRI, which can drive mast-cell activation (see Fig. 11.2). Tests for anti-nuclear antibody (ANA) and anti-double-stranded DNA (anti-dsDNA) antibody are useful screens for the autoimmune disease systemic lupus erythematosus, in which hives can arise from cutaneous vasculitis and/or immune-complex-mediated mast-cell activation. The serum complement CH_{50} assay tests for total complement activity. Complement consumption can occur in immune-complex-mediated conditions and is associated with a decreased CH_{50} value. Serologic testing can be used to rule out viral hepatitis, and both serological tests and stool analysis can be used to search for evidence of parasite infestation. Finally, a number of physical challenges can be applied to look for physical urticaria, for instance the ice-cube test in which the application of an ice cube to the forearm for 15 minutes gives rise to a large wheal and flare.

Several pharmacologic options are available to manage urticaria (Fig. 11.5). Antihistamines are the mainstay of therapy (see Case 3). In urticaria, histamine released by activated dermal mast cells binds to the type 1 histamine receptor (H1R) on the endothelium of small dermal blood vessels, leading to increased vascular permeability with plasma extravasation and localized edema. Diphenhydramine and hydroxyzine are two commonly used first-generation H1R-blocking antihistamines. The first-generation antihistamines are very effective in treating urticaria, but they cross the blood–brain barrier and can cause drowsiness and sedation because of their effects on H1R in the central nervous system (CNS). In children, there is often a paradoxical stimulant effect of H1R blockers, leading to excitement and sometimes agitation. Second-generation H1R blockers (loratidine, cetirizine, and fexofenadine) which do not cross into the CNS, and thus have little sedative effect, are now the first choice for treatment of urticaria.

In cases in which H1R blockers do not produce sufficient relief of symptoms, antagonists that target type 2 histamine receptors (H2R), such as ranitidine or cimeditine, are typically added. However, the role of H2R in urticaria has not been well characterized, and there is minimal, if any, additional benefit of

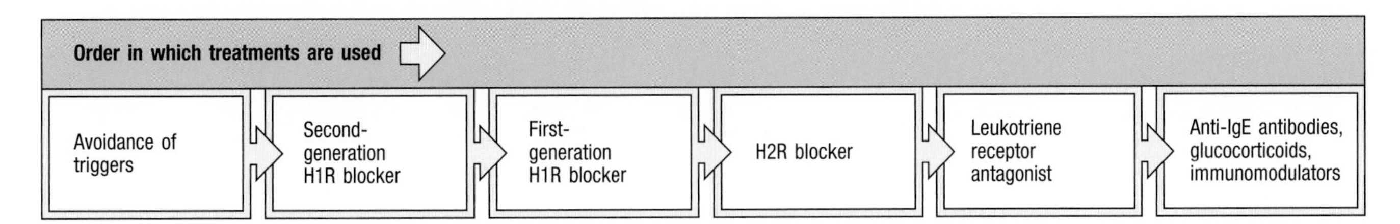

Fig. 11.5 Stepwise approach to urticaria therapy. In patients in whom allergy testing or medical history has established an allergic trigger for hives, avoidance of the trigger is advised. When the trigger is unknown, or avoidance is impracticable, most patients achieve very good control of their urticaria using well-tolerated second-generation H1R blockers such as cetirizine or loratadine. A first-generation H1R blocker can be added if needed, but this therapy can be associated with unacceptable side effects of sedation. When these standard approaches prove ineffective, additional options include the addition of an H2R blocker (ranitidine or cimetidine), a leukotriene receptor antagonist (montelukast or zafirlukast) and, finally, immunomodulatory agents such as glucocorticoids, hydroxychloroquine, and cyclosporin, and anti-IgE monoclonal antibody that removes IgE from the circulation and decreases its binding to mast cells.

these agents. Typically, patients are advised to take antihistamines as needed when hives occur, but the medications can also be used prophylactically in cases in which the hives are severe enough to interfere with normal daily activities.

In severe refractory chronic urticaria, other medications that target mast-cell and immune activation can be used, such as leukotriene receptor antagonists and immunosuppressive medications including corticosteroids, hydroxychloroquine, or cyclosporin. Although drug therapy is generally quite effective, urticaria can have a significant impact on patients' quality of life and be a source of much frustration when the underlying cause cannot be identified. Thus, reassurance and education by an informed clinician can greatly benefit patients and their families.

Questions.

1 *Why did Fitzwilliam have hives when he ate peanut butter at 2 years of age? Why were his symptoms different more recently when he petted the cat?*

2 *Epinephrine (often known by its alternative name, adrenaline) is commonly used as a treatment for acute anaphylaxis because of its effects on the peripheral vasculature and cardiac contractility. How would epinephrine help to resolve hives? Why is epinephrine not used to treat isolated cases of urticaria?*

3 *Hives can occur on any skin surface, but occur less often on the palms of the hands and soles of the feet, and more often in areas exposed to pressure (such as around the waistband of clothing). Why might this be the case?*

4 *How would you test for cold-induced urticaria?*

CASE 12 IgE-Mediated Food Allergy

A severe reaction after eating.

The normal physiologic immune response to foods is the induction of tolerance (Fig. 12.1). Small quantities of undigested food proteins are transported across the intestinal epithelium to the mucosa, where antigen-presenting cells process and present them to naive T cells, resulting in T-cell priming and expansion. Under homeostatic conditions, regulatory T cells (T_{reg} cells) are preferentially induced. T_{reg} cells actively suppress helper T-cell (T_H) responses in an antigen-specific manner and are characterized by their expression of the transcription factor Foxp3. Boys harboring mutations of the *FOXP3* gene (which is on the X chromosome) exhibit severe food allergy, along with a number of other features of immune dysregulation.

In some individuals, a combination of genetic predisposition and environmental factors results in the redirection of immune responses to food allergens from T_{reg}-enforced tolerance to T-cell-driven IgE-mediated hypersensitivity. In these subjects, the T-cell response to food proteins is dominated by T_H2 cells. These produce the key effector cytokines of allergy—IL-4, IL-5, and IL-13—which support the induction of IgE responses and mast-cell expansion as well as regulating intestinal epithelial functions. Ligands for innate immune-system receptors, including Toll-like receptors (TLR), regulate the function of antigen-presenting cells in the intestine and influence their induction of T_{reg}-cell versus T_H2-cell responses. For this reason, the composition of the intestinal microflora may have a critical role in determining the food allergy response. It has been shown that exposure to pets, farmyard animals, or raw cow's milk (all potential sources of microbiota containing TLR ligands) during childhood protects against food allergy.

Intestinal permeability is an important factor in the development of food allergy. Both nonspecific paracellular penetration of intact food antigens through the spaces between epithelial cells and antigen-specific transcellular movement, facilitated by preexisting IgE antibodies bound to the low-affinity IgE receptor, CD23, are probably enhanced in allergic individuals. This results in greater absorption of food allergens into subepithelial intestinal mucosal tissues, facilitating both immune sensitization and the elicitation of allergic

This case was prepared by Hans Oettgen, MD, PhD, and Raif Geha, MD, in collaboration with Lisa Bartnikas, MD.

Topics bearing on this case:

- IgE antibodies
- Immune regulation
- Regulatory T cells
- Mast cells

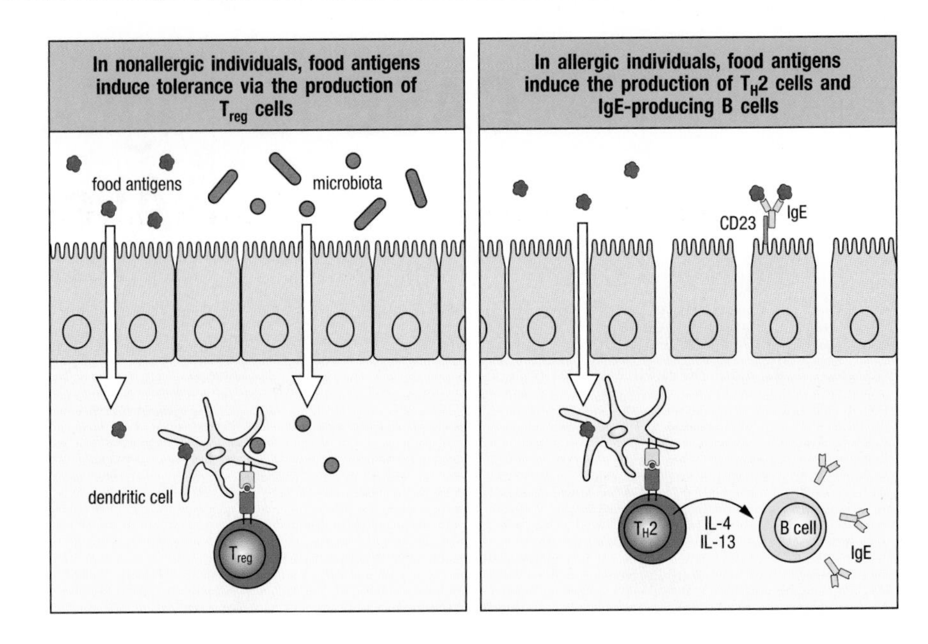

Fig. 12.1 Allergic sensitization to foods. In nonallergic individuals, food proteins that cross the gut epithelium are taken up by antigen-presenting cells (APCs) in the lamina propria that drive the differentiation of regulatory T cells (T_{reg}) rather than of other T-cell subsets such as T_H2 (left panel). The interaction of molecules on the gut microbiota with innate immune receptors on APCs may be important in maintaining the pressure to induce T_{reg} cells. In individuals with a food allergy, several processes converge to induce an IgE response (right panel). Allergic subjects have increased intestinal permeability, allowing intact macromolecules to penetrate the gut wall more easily through gaps between the intestinal epithelial cells. In addition, once some food-specific IgE antibodies are present, these can facilitate the transcellular uptake of antigen by the low-affinity IgE receptor CD23, located on the apical surface of the intestinal cells. Both processes make intact high-molecular-weight antigens, such as food proteins, more available to APCs. Specific gut microflora in individuals susceptible to food allergy may affect the APCs so as to favor T_H2-cell induction. These activated T_H2 cells make cytokines, including IL-4 and IL-13, that stimulate B cells to synthesize food-antigen-specific IgE.

responses by IgE-bearing intestinal mast cells. Differences in the numbers, tissue distribution, and activation threshold of mast cells dictated by the same genetic factors that lead to atopy are also likely determinants of the food allergy response.

Milk, soy, egg, wheat, peanuts, tree nuts, fish, and shellfish account for 90% of allergic reactions to food. The allergens contained in these foods are water-soluble glycoproteins 10–70 kDa in size, and are relatively resistant to degradation by heat, acid, and proteases. A single food can contain multiple allergenic proteins, each bearing many epitopes recognizable by IgE antibodies. Up to 80% of children with milk or egg allergy can tolerate well-baked forms of the protein (for example in muffins or cookies) but will develop allergic reactions to unbaked forms (for example a glass of milk or scrambled eggs). Children unable to tolerate any form of milk or egg have IgE antibodies recognizing heat-stable linear epitopes. In contrast, the immune response of children who tolerate well-baked forms of milk or egg is directed against conformational epitopes, which are lost when the protein unfolds on heating (Fig. 12.2).

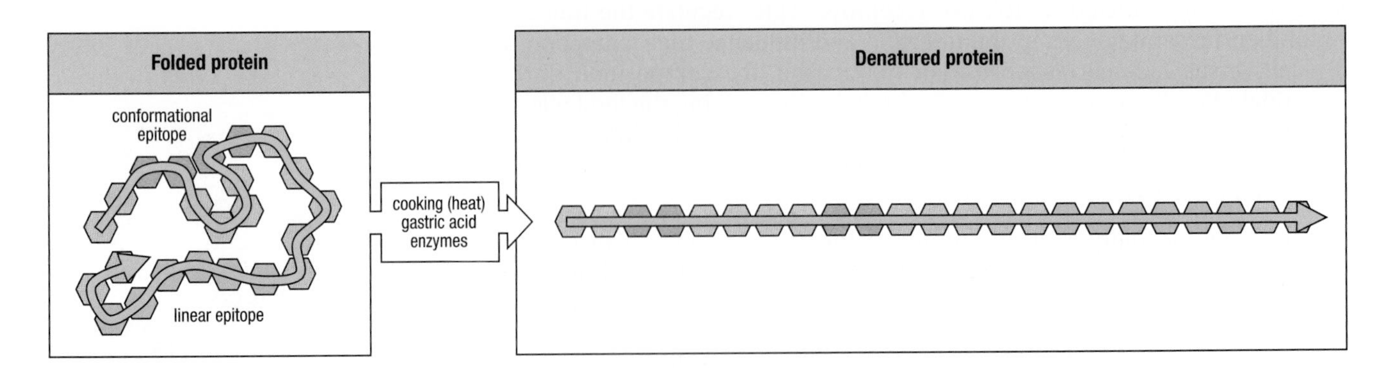

Fig. 12.2 A single protein can generate a variety of epitopes. The native folded form of a protein carries a variety of different epitopes, including conformational epitopes and linear epitopes. The denaturation of the protein by cooking, followed by digestion, leads to the loss of conformational epitopes. Individuals with antibodies recognizing the conformational epitopes can therefore tolerate well-cooked but not uncooked forms of the food. This phenomenon can be seen in milk and egg allergies.

The case of Gregor Samsa: a severe reaction only to certain forms of egg.

Gregor Samsa was brought to the Children's Hospital Emergency Department at 2 years old because of a severe allergic reaction. Gregor had been healthy until 2 months of age, when he developed atopic dermatitis. At 6 months, his mother fed him scrambled eggs and after just a few bites he spat them out, refusing to eat more. Subsequently, he began eating muffins, cookies, and other egg-containing baked goods without any problem.

On the day he was taken ill, Gregor had been to a family brunch and sat next to his grandfather, who was eating an omelet. Gregor did not eat the omelet but his grandfather used his fork to feed Gregor some fruit from his plate. Within minutes, Gregor developed hives, vomiting, sneezing, and itchy red eyes. He was brought to the Emergency Room, where he was treated with intramuscular epinephrine, which resolved his symptoms. He was observed for several hours for possible recurrence of symptoms and then discharged with a prescription for injectable epinephrine.

2-year-old boy with anaphylaxis.
Positive skin and blood tests for egg-specific IgE.

Gregor was also referred for further allergy testing. After a careful review of his history, skin-prick testing for egg allergy was performed and was positive, with a 25-mm wheal and a 40-mm flare. Total serum IgE was elevated, at 500 IU ml^{-1} (normal 0–30 IU ml^{-1}). Specific IgE to total egg-white proteins was elevated at 8.65 IU ml^{-1} (normal less than 0.35 IU ml^{-1}) and specific IgE to ovomucoid, a heat-stable protein in eggs, was undetectable (at less than 0.35 IU ml^{-1}) (Fig. 12.3). Gregor was instructed to avoid uncooked egg but to continue eating egg in well-baked goods, and the family was advised to carry two self-injectable syringes of epinephrine at all times for treatment in case of a severe allergic reaction.

When he turned 7 years old, Gregor returned to the allergy clinic for follow up. Skin-prick testing for egg allergy was negative and the IgE specific for egg had dropped to 0.5 IU ml^{-1}, with specific IgE to ovomucoid again undetectable at less than 0.35 IU ml^{-1}. His allergist recommended a food challenge to uncooked egg to see whether he might have outgrown his egg allergy. He successfully ate incrementally increasing amounts

Levels of specific IgE antibodies indicating a food allergy vary for different foods

Food	Positive predictive value (PPV) in IU ml^{-1}		Negative predictive value (NPV) in IU ml^{-1}	
	> 95% probability	> 90% probability	> 95% probability	> 90% probability
Milk	32	23	0.8	1
Soy	best PPV ascertained = 50% at 65 IU ml^{-1}		2	5
Egg	6	2	–	0.6
Wheat	best PPV ascertained = 75% at 100 IU ml^{-1}		5	79
Peanut	15	9	best NPV ascertained = 85% at 0.35 IU ml^{-1}	
Fish	20	9.5	0.9	5

Fig. 12.3 Levels of specific IgE antibodies indicating a food allergy vary for different foods. Studies of individuals who passed or failed food challenges have yielded values for food-specific serum IgE that predict an allergic reaction to a particular food. The positive predictive value (PPV) gives the risk of being allergic and failing a food challenge, while the negative predictive value (NPV) gives the risk of not being allergic and passing a food challenge. Note that predictive values vary widely for different foods. An NPV of more than 95% for egg could not be established.

Passed egg challenge:
egg allergy outgrown!

of scrambled egg, totaling 10 g of protein, without reaction, so Gregor was considered to have outgrown his egg allergy. He was subsequently able to consume egg in all forms without reaction.

IgE-mediated food allergy.

Currently, 6–8% of American children and 3–4% of adults have IgE-mediated food allergies, and the incidence is on the rise. Of children with moderate to severe atopic dermatitis, 30–40% have IgE-mediated food allergies. Food allergens are the most common triggers of anaphylaxis, and the most feared complication of food allergy is death from anaphylaxis. The rapid rise of food allergies indicates that environmental factors rather than genetic predisposition are driving the trend. The potential influences of antibiotic use, childhood immunization, administration of antacids (which can alter allergen structure by preventing gastric acid hydrolysis), deficiency of vitamins D and E, duration of breastfeeding, environmental effects on intestinal microflora, and age of introduction of certain foods are all under investigation.

A diagnosis of food allergy is based on a history of typical symptoms of immediate hypersensitivity occurring reproducibly upon exposure to a

Fig. 12.4 Mechanisms of food allergy. Food allergies can be mediated by IgE, by T cells, or by both. Depending on the underlying immunological mechanism, particular foods may be involved and the allergic symptoms displayed may be different.

Immunopathology	Disorder	Key features	Typical age	Most common causal foods
IgE-mediated	Urticaria, angioedema, and anaphylaxis	Rapidly progressive, potentially life-threatening multi-system reaction in the case of anaphylaxis	Any	Any, but most commonly milk, soy, egg, wheat, peanut, tree nuts, fish, and shellfish
	Oral allergy syndrome	Itching and swelling limited to oral mucosa as a result of cross-reactivity between pollen and food proteins. Around 7% progress beyond the mouth and 1–2% have anaphylaxis	Onset after pollen allergy established. More common in older children and adults	Raw fruits/vegetables. Cooked forms are tolerated
	Food-associated, exercise-induced anaphylaxis	Food triggers anaphylaxis only if ingestion is followed by exercise within 2–4 hours	Onset more common in older children and adults	Wheat, celery
T-cell-mediated	Food-protein-induced enterocolitis syndrome	Chronic exposure: vomiting, diarrhea, and poor growth. Acute exposure: vomiting, diarrhea, and hypotension 2 hours after ingestion	Infants	Milk, soy, rice, oats
	Food protein proctitis	Bloody stools in infants	Infants	Milk, soy
Mixed IgE- and T-cell-mediated	Atopic dermatitis (see Case 5)	Associated with food in 30–40% of children with moderate to severe rash	Infant > child > adult	Major allergens, especially egg and wheat
	Eosinophilic esophagitis (see Case 13)	Pain and difficulty swallowing, food impaction	Any	Any, but most commonly milk, soy, egg, wheat, peanut, tree nuts, fish, and shellfish

food allergen. Symptoms of IgE-mediated reactions are those of mast-cell and basophil mediator release and can affect a number of organ systems including skin (hives, angioedema), respiratory (stridor, cough, wheezing), gastrointestinal (nausea, vomiting, abdominal pain, diarrhea), cardiovascular (hypotension), and central nervous system (loss of consciousness, feeling of impending doom). Symptoms occur within minutes and up to 2 hours after ingestion, and resolve within several hours. Some patients exhibit delayed and chronic reactions to foods, which are driven by IgE-independent immunological mechanisms. Although this case focuses on IgE-mediated food allergy, the most common, and also the most serious, form of sensitivity to food, it is important to be aware of the existence of these other immunologically mediated food reactions (Fig. 12.4). They include T-cell-mediated enterocolitis syndrome, food-induced atopic dermatitis and eosinophilic esophagitis (see Case 13). An unusual delayed form of IgE-mediated food sensitivity has also been reported, in which anaphylaxis to mammalian meat is triggered by IgE directed against the carbohydrate galactose-α-1,3-galactose (α-Gal) rather than a protein epitope, leading to symptoms 3–6 hours after ingestion.

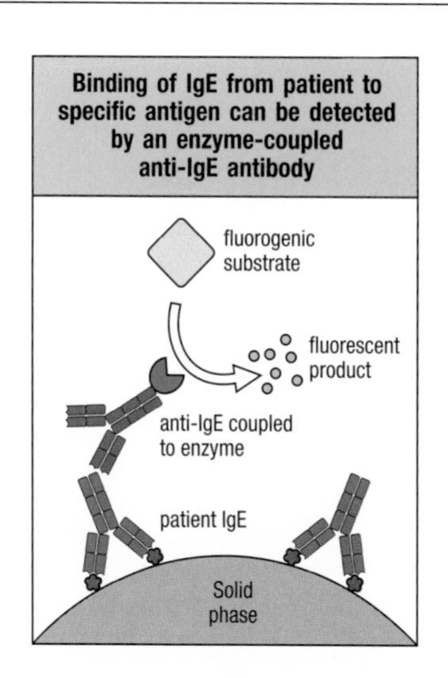
Binding of IgE from patient to specific antigen can be detected by an enzyme-coupled anti-IgE antibody
fluorogenic substrate
fluorescent product
anti-IgE coupled to enzyme
patient IgE
Solid phase

Fig. 12.5 Principles of the fluorenzyme immunoassay (FEIA) for antigen-specific serum IgE. The allergen of interest is coated onto the surface of a cellulose solid phase. Serum samples from the patient are then applied and any specific antibodies present will bind tightly to their target antigen. Unbound antibody is removed by washing. Anti-IgE antibody that has been chemically linked to an enzyme is then added and, after further washing, a substrate for the enzyme is added. This produces a measurable fluorescent product after enzymatic cleavage.

Testing for the presence of antigen-specific IgE can be done by laboratory tests or by *in vivo* skin-prick testing. Serum assays for allergen-specific IgE are reported on a scale of less than 0.35 to more than 100 IU ml^{-1} (Fig. 12.5). Skin-prick testing is performed by gently scratching the surface of the skin and then applying a drop of an aqueous extract of the allergen. In the presence of allergen-specific IgE, cutaneous mast cells are activated and a hive surrounded by a flat erythematous flare appears within 15 minutes. A wheal of diameter 3 mm greater than the negative saline control is considered positive. Allergen-specific IgE levels and the results of skin-prick testing are used to determine the likelihood of clinical allergy to food (see Fig. 12.3). It is crucial that all laboratory testing be interpreted in the context of the clinical history, because some patients have detectable levels of antigen-specific IgE, yet tolerate a food. The fact that individual patients with comparable levels of allergen-specific IgE to a given food can either tolerate the food without incident or manifest a severe reaction upon ingestion indicates that other factors, including intestinal permeability and mast-cell phenotype, must be important determinants of allergic reactions. Because these are poorly defined and not readily evaluated, it is often necessary to perform food challenges in a clinical setting to diagnose food allergy (or the lack thereof) with certainty.

Currently, no cure for food allergy has been approved for use in routine clinical practice. The focus is on avoidance of food allergens, recognition of early signs of anaphylaxis, and training on the use of self-injectable epinephrine in the event of a reaction. Contaminated cooking equipment or utensils can result in the inadvertent ingestion of allergen, as we saw with Gregor. Unfortunately, at least 15% of food-allergic patients have reactions due to accidental ingestions each year. Emergency preparedness with access to two doses of epinephrine is crucial, because up to 12% of patients with food allergy-induced anaphylaxis will require a second dose of epinephrine. Treatment with antihistamines, such as diphenhydramine, may alleviate mild symptoms such as itching related to mediator release, but will not prevent the progression of anaphylaxis.

Children can outgrow their food allergies. The rate of resolution of allergy is specific to each food. Approximately 60% of children with food allergy to milk, egg, soy, and wheat will eventually tolerate these foods. However, only 20% will outgrow peanut allergy and 10% will outgrow tree nut allergy. Food allergy that develops in adult life tends to persist. Children who tolerate well-baked forms of milk or egg, as Gregor does, have been observed to outgrow these respective allergies faster than those who are not tolerant to well-baked forms. Eating the foods in well-baked forms may directly accelerate resolution of the allergies, possibly via the induction of allergen-specific T_{reg} cells.

Questions.

1 *Why did Gregor develop reactions to uncooked egg but not well-baked egg?*

2 *What vaccines should egg-allergic patients avoid?*

3 *Should children with egg allergy avoid eating chicken?*

4 *Even though Gregor still tested positive for specific IgE against egg, why was it a reasonable decision to proceed with a food challenge to egg?*

5 *Gregor's mother is now pregnant and asks for your opinion regarding breast versus bottle feeding and when to introduce specific foods to the new baby. What do you tell the family?*

6 *Can you think of any immunodeficiencies associated with increased risk of food allergies?*

CASE 13 Eosinophilic Esophagitis

A difficulty in swallowing caused by a chronic allergic reaction to food.

Immune responses to food allergens can be mediated by a range of effector mechanisms, broadly divided into IgE- and non-IgE-mediated responses. The classic manifestation of IgE-mediated food allergy is anaphylaxis, in which food ingestion leads to rapid mast-cell activation and multisystem symptoms (see Cases 1 and 12). Non-IgE-mediated food hypersensitivities are dominated by cellular immune responses, driven by T cells specific for food antigens, and are responsible for a range of disorders including milk-protein proctocolitis of infancy and food-protein-induced enterocolitis syndrome. In such cell-mediated responses, food antigens are initially taken up and processed by antigen-presenting cells residing in the intestinal mucosa. These cells move to regional lymphoid tissues, including Peyer's patches in the gastrointestinal mucosa, which are organized collections of T cells and B cells similar to the lymphoid follicles and T-cell areas in the spleen or lymph nodes. There, antigen-presenting cells present peptide fragments of food antigens in association with MHC class II molecules to T cells. When the antigen receptor of a T cell exposed to an antigen-presenting cell recognizes and binds the peptide:MHC complex with appropriate co-stimulatory signals, the T cell undergoes activation and clonal expansion. Depending upon the local milieu of cytokines and co-stimulatory molecules, the resulting T-cell response can take on a range of effector phenotypes, resulting in a chronic inflammatory response induced by various mechanisms. Chronic inflammation involving a T-cell response can also follow on from an IgE-initiated allergic response as, for example, in the case of chronic asthma (see Case 2) (Fig. 13.1).

The eosinophil has a central role in several chronic allergic conditions. The contents of the eosinophil granules are released on activation and cause inflammation and tissue damage. Eosinophils accumulate in the lungs of patients with asthma, the skin of individuals with atopic dermatitis (see Case 5) during acute flares of their rashes, and in the nasal mucosa of patients with allergic rhinitis (see Case 3). As we see in the case of eosinophilic esophagitis, they are recruited into the esophagus, causing an inflammation driven by a food-specific adaptive immune response.

Topics bearing on this case:

- Eosinophils
- Allergic reactions to food
- Delayed-type (type IV) hypersensitivity reactions

This case was prepared by Hans Oettgen, MD, PhD, and Raif Geha, MD, in collaboration with John Lee, MD.

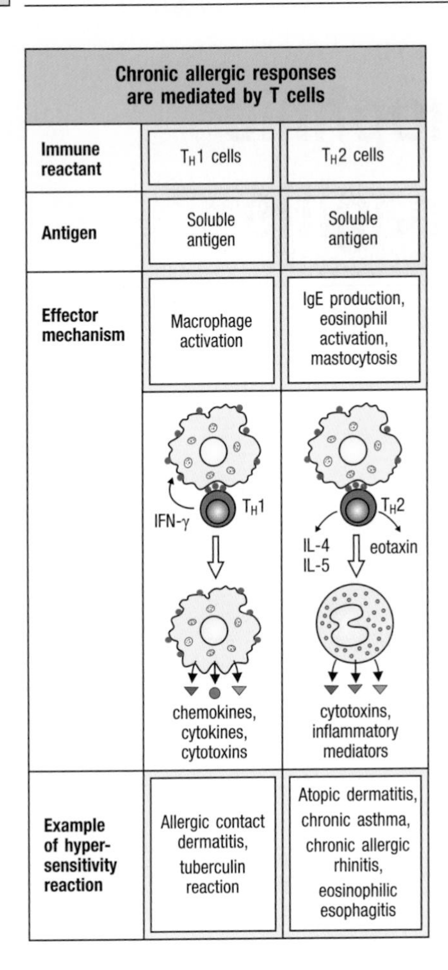

Chronic allergic responses are mediated by T cells
Immune reactant
TH1 cells
TH2 cells
Antigen
Soluble antigen
Soluble antigen
Effector mechanism
Macrophage activation
IgE production, eosinophil activation, mastocytosis
IFN-γ
TH1
TH2
IL-4
IL-5
eotaxin
chemokines, cytokines, cytotoxins
cytotoxins, inflammatory mediators
Example of hyper-sensitivity reaction
Allergic contact dermatitis, tuberculin reaction
Atopic dermatitis, chronic asthma, chronic allergic rhinitis, eosinophilic esophagitis

Fig. 13.1 Cell-mediated allergic reactions are T-cell mediated.
In some chronic allergic reactions, tissue damage is caused by the activation of macrophages by T_H1 cells, which results in an inflammatory response. In another type, damage is caused by the activation by T_H2 cells of allergic inflammatory responses in which eosinophils predominate; this type of response often follows an initial IgE-mediated allergic response. Eosinophilic esophagitis is caused primarily by T_H2 activation.

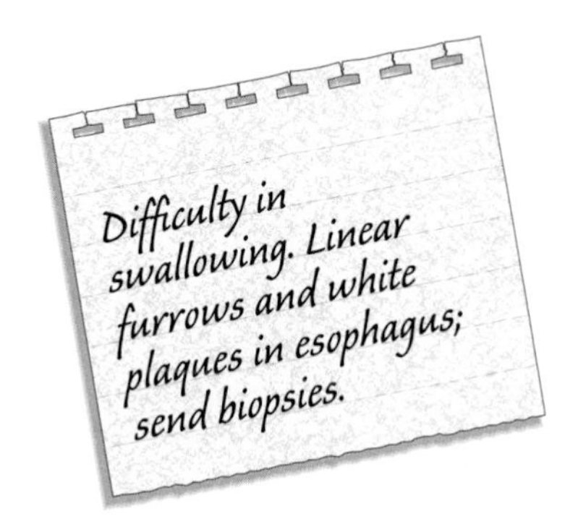

Difficulty in swallowing. Linear furrows and white plaques in esophagus; send biopsies.

The case of Buck Mulligan: a 15-year-old boy with a food impaction.

Buck Mulligan was seen at the emergency room of the Children's Hospital. He was eating dinner at home with his family when a piece of steak he had just swallowed felt as if it was lodged in his throat, causing him considerable amount of discomfort in the chest. He tried to eat some more but this resulted in vomiting of undigested food. When he tried to take a few sips of water, this also came back up. After a few hours without the symptoms improving, his family brought him to the emergency department.

A barium swallow test was performed which revealed an obstruction in his lower esophagus preventing the passage of radiocontrast material. He subsequently underwent endoscopy, during which the gastroenterologist noted the presence of linear furrows and white plaques lining his esophagus. Multiple esophageal biopsies were taken. A piece of steak was retrieved from the lower third of his esophagus. Buck was started on the proton-pump inhibitor omeprazole for empiric therapy of acid-reflux disease and was discharged home the following day.

Histopathologic examination of the esophageal biopsies revealed numerous eosinophils (more than 100 per high-power field) within the epithelium of the esophagus. For several weeks afterwards, Buck had no further food-impaction episodes but had frequent episodes of dysphagia (difficulty in swallowing). A repeat endoscopy after the course of acid-reflux therapy showed a persistent eosinophilic infiltration in the esophageal biopsies, confirming the diagnosis of eosinophilic esophagitis.

After eosinophilic esophagitis was diagnosed, Buck's gastroenterologist started him on swallowed fluticasone therapy. Within a few weeks, Buck noticed significantly fewer episodes of dysphagia. After 2 months on fluticasone, he had complete resolution of symptoms. However, one day he developed acute throat pain. He had a repeat endoscopy, which showed resolution of the eosinophilia, but microscopic examination of esophageal brushings revealed pseudo-hyphae indicative of the yeast *Candida*. Buck was treated with a course of fluconazole, which resulted in clearance of his symptoms. He was then referred to the allergy clinic for further evaluation.

Buck had a history of allergic rhinitis for several years, with particularly bad symptoms during the spring pollen months. He had also had intermittent symptoms of mild asthma that were easily controlled using the inhaled bronchodilator albuterol (a β-adrenergic agent). Buck reported having experienced a transient sensation of food getting stuck in his throat for a few seconds for the past 4 or 5 years, typically when eating bread or meat. His family had attributed this symptom to his eating too quickly. For the most part he was able to compensate by chewing more carefully and taking frequent sips of water with meals. His dysphagia seemed to be more frequent when he was symptomatic from his spring pollen allergies.

During his visit to the allergy clinic, Buck had skin-prick tests for a panel of foods and environmental aeroallergens. He had positive reactions to milk, egg, and chicken, as well as to some environmental allergens, including tree pollens, ragweed pollen, and dust mites. A complete blood count revealed a mild increase in his peripheral eosinophils (875 cells μl^{-1}; normal range 0–500) but otherwise was normal. Buck's serum IgE was mildly increased at 550 IU ml^{-1} (normal range 0–150 IU ml^{-1}). He subsequently had atopy patch testing to evaluate cell-mediated hypersensitivity to foods. The test patches were removed 48 hours after placement, and the results were read at 72 hours. Erythema and papules were noted at the site of the patch tests with oat and soy (2+ reaction).

Buck was put on a specific food elimination diet avoiding milk, egg, chicken, soy, and oats while tapering off his swallowed fluticasone. He had stopped his acid-reflux

medications. After 2 months, fluticasone was stopped and he continued to follow his restricted diet without any recurrence of symptoms. A repeat endoscopy showed a normal healthy esophagus with no return of eosinophilia in the biopsies. One year later, Buck has adjusted well to his dietary restriction and remains symptom free.

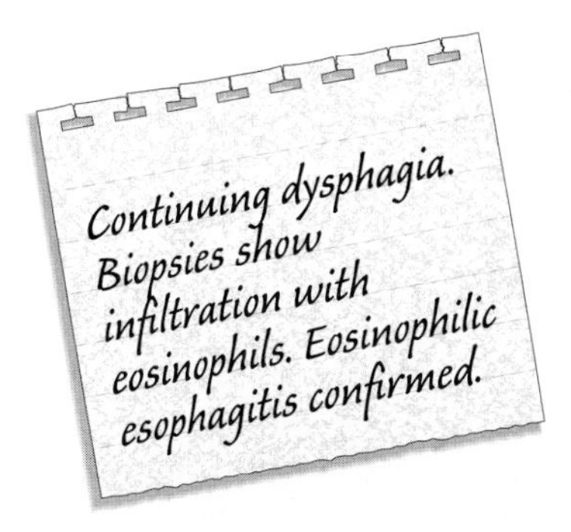

Continuing dysphagia.
Biopsies show
infiltration with
eosinophils. Eosinophilic
esophagitis confirmed.

Eosinophilic esophagitis.

Eosinophilic esophagitis is a chronic allergen-driven disorder in which eosinophils are found in large numbers in the inflamed esophageal tissue. The prevalence of this disease has increased severalfold in the past decade, and currently 1 in 10,000 children are affected. It is more often seen in Caucasians and is more common in males than females, at a ratio of 3:1. More than half the patients also have other atopic conditions. Diagnosis is based on a clinicopathologic constellation of symptoms, histologic features of esophageal biopsies, and exclusion of other disorders.

Eosinophilic esophagitis, like atopic dermatitis (see Case 5), spans the spectrum of IgE- and non-IgE-mediated mechanisms. Patients with this disorder have evidence of both IgE-mediated and cellular immunological responses to food allergens. Skin-prick testing or ELISA assays for antigen-specific IgE demonstrate the presence of food-specific IgE in patients with eosinophilic esophagitis, while atopy patch testing provides evidence for the presence of food-specific T cells capable of driving tissue inflammation. In addition to T_H1 and T_H2 cells, other immune effector cells including eosinophils, basophils, macrophages, NK cells, invariant NKT cells, and cytotoxic T cells can also participate in the pathophysiology of IgE-independent food sensitivity.

Symptoms of eosinophilic esophagitis can vary depending on the age of presentation. Infants show feeding refusal, poor growth, or symptoms of gastroesophageal reflux (vomiting, feeding difficulty, irritability after feeds). Young children can complain of vomiting and abdominal pain. Dysphagia and food impaction are more commonly seen in adolescents and adults. Some patients with eosinophilic esophagitis can have atypical symptoms such as chest pain or cough, or can be asymptomatic. Eosinophilic esophagitis and gastroesophageal reflux disease have overlapping features. Both can present with esophageal eosinophilia and symptoms of heartburn; it is therefore essential to distinguish between these two conditions. This is done by obtaining esophageal biopsies while on a proton-pump inhibitor (such as omeprazole). Anti-reflux medications should relieve symptoms and resolve the tissue eosinophilia in the reflux disease but not in eosinophilic esophagitis.

Endoscopy can reveal several features characteristic of eosinophilic esophagitis. However, none of these is pathognomonic (that is, so characteristic of the disease that it constitutes a diagnosis). Common findings include linear furrows, eosinophilic abscesses that appear as white specks (plaques), and pale friable mucosa with decreased vascularity (Fig. 13.2). Occasionally, circumferential rings (known as trachealization of the esophagus or a feline esophagus), esophageal strictures, or a food bolus are observed. Radiographic findings suggestive of eosinophilic esophagitis include narrowing of large segments of the esophagus, Schatzki rings (discrete narrowing of the lower esophagus), or a food-bolus impaction on barium swallow (Fig. 13.3).

Esophageal biopsies are required to make the diagnosis of eosinophilic esophagitis, with a minimum of 15 eosinophils seen per high-power field (Fig. 13.4). In addition, basal-cell hyperplasia and epithelial thickening are also commonly observed and support the diagnosis. Mast cells, basophils,

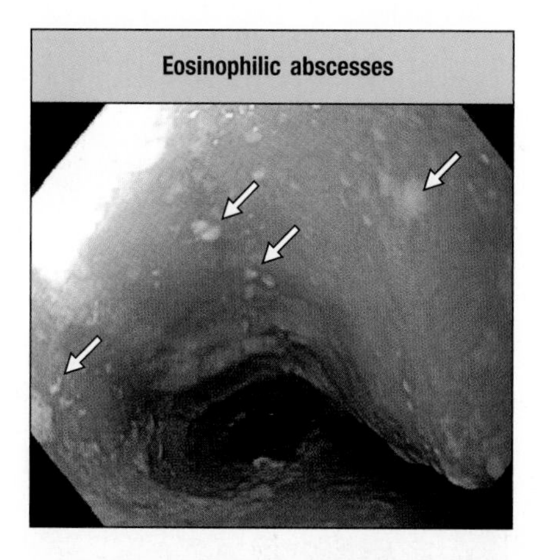

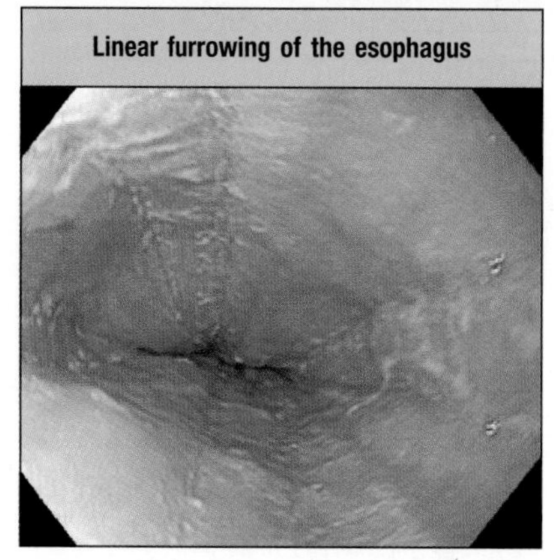

Fig. 13.2 Esophageal endoscopy in eosinophilic esophagitis. Multiple eosinophilic abscesses are seen as white specks (top panel). Linear furrowing of the esophagus (bottom panel).

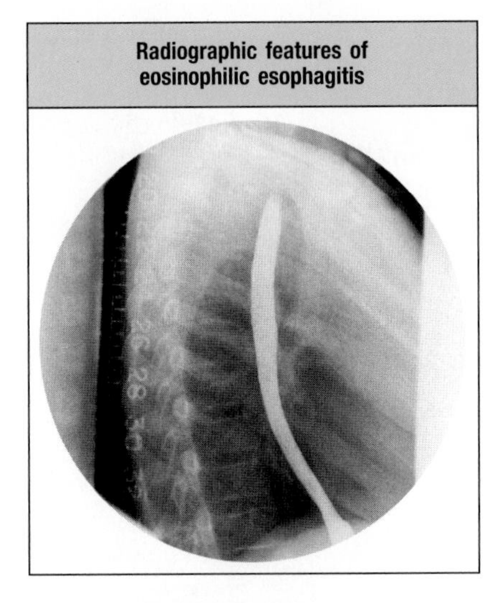
Radiographic features of eosinophilic esophagitis

Fig. 13.3 Radiographic features of eosinophilic esophagitis. Smooth esophageal narrowing can be observed at the distal third of the esophagus.

and lymphocytes may also be present in increased numbers in esophageal biopsies.

In children, eosinophilic esophagitis is driven primarily by food allergy, so dietary therapy can be an effective means of controlling inflammation. Three major strategies are used. The most effective, with nearly 100% resolution rates, is the implementation of an elemental diet. This involves placing patients on an amino-acid-based hypoallergenic formula with exclusion of all other foods. The very significant negative social and nutritional aspects of being on a formula-only diet have, however, limited its widespread use. An alternative is an empiric elimination diet in which the most common food triggers associated with the disease are eliminated. A typical diet entails the avoidance of milk, egg, wheat, soy, peanuts, tree nuts, and seafood. A third strategy is the focused elimination of foods on the basis of the results of allergy testing, as in Buck's treatment.

Skin-prick testing and specific IgE measurements are both used to screen for possible food triggers. However, because eosinophilic esophagitis is due to a mixture of IgE- and non-IgE-mediated hypersensitivities, these tests may not identify all causative foods, so atopy patch testing has been used additionally to screen for non-IgE-mediated food allergies. In patch testing, the suspected allergen is applied to the skin for a period of 48 hours. During this time, dendritic cells in the skin process and present antigens to circulating T cells, which are constantly surveying tissue sites for the presence of foreign antigens. If specific T cells are present, a local inflammatory response ensues at the site of the patch test, resulting in local erythema and papules that evolve over 24–48 hours after removal of the patch (Fig. 13.5).

Topical steroids have been used to treat the inflammation of eosinophilic esophagitis. Glucocorticoid preparations commonly used for asthma via inhalers are given, along with instructions to swallow rather than inhale the medicine upon activation of the device. Steroids can relieve the eosinophilia and reverse the fibrosis seen in the disease.

Eosinophil infiltration is the hallmark of this disease. In a normal healthy esophagus no eosinophils are found. However, in eosinophilic esophagitis, eotaxin-3, a potent chemotactic factor for eosinophils, and IL-5, a cytokine important in eosinophilopoiesis and trafficking, are both highly expressed in the esophageal epithelium and are likely to have a major role in driving the recruitment of circulating eosinophils into the tissue. IL-13, a major $T_{H}2$ cytokine, is also present at elevated levels within the esophagus of eosinophilic esophagitis patients and is likely to have a central role in the esophageal remodeling and epithelial hyperplasia associated with the disorder. Another cytokine strongly associated with the disease is thymic stromal lymphopoietin, which is a master regulator for allergic immune responses. It is produced by epithelial cells and can act on CD4 T cells to divert them to an allergic $T_{H}2$ phenotype. Under the influence of this cytokine, dendritic cells produce chemokines, including CCL17 and CCL22, which attract CD4 $T_{H}2$ cells.

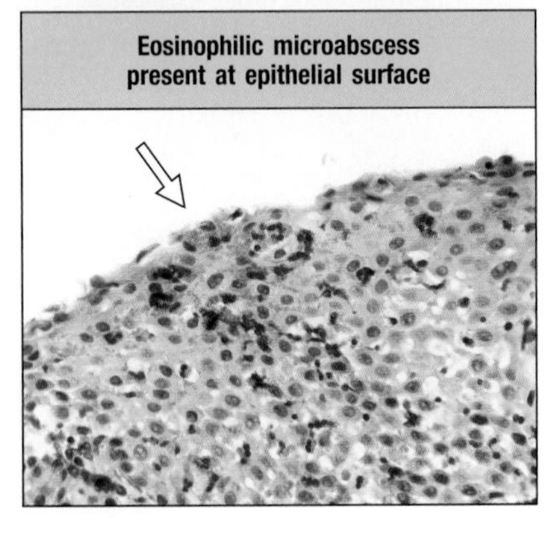
Eosinophils infiltrating epithelium of esophagus
Eosinophilic microabscess present at epithelial surface

Fig. 13.4 Esophageal biopsies in eosinophilic esophagitis. Numerous eosinophils are seen infiltrating the epithelium of the esophagus (upper panel). In the lower panel, an organized eosinophilic microabscess is present at the epithelial surface (arrowed).

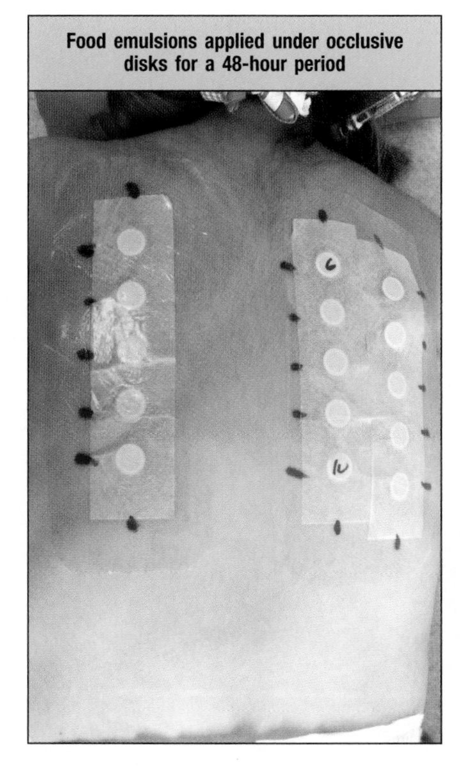

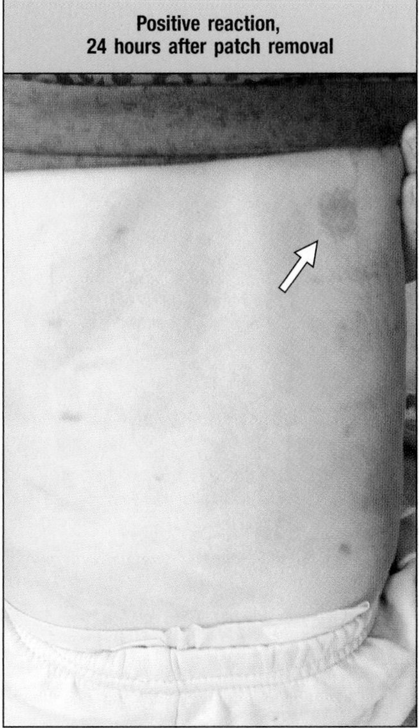

Fig. 13.5 Food allergen patch testing. Food emulsions are applied to the patient's back under occlusive disks for a 48-hour period, as seen in the left panel. A positive reaction, 24 hours after patch removal, is seen at the upper right quadrant of the back in the right panel.

Questions.

1 *Amino-acid-based formulas have been shown to be very effective in the management of eosinophilic esophagitis. Why are these elemental formulas considered to be hypoallergenic?*

2 *A neutralizing monoclonal antibody against IL-5 has been evaluated for the treatment of eosinophilic esophagitis, but has not resulted in complete remission of the disease. Explain why anti-IL-5 may not be a fully effective therapy.*

3 *Buck noticed that his esophageal symptoms also flared in the springtime along with his allergic rhinitis. Explain why this occurred.*

4 *Why does the food to which Buck was allergic not result in symptoms of anaphylaxis?*

CASE 14 Allergic Bronchopulmonary Aspergillosis

An allergic reaction initiated by an innate immune response.

Aspergillus species are spore-forming fungi ubiquitous in the environment. When they are encountered by humans, they can induce diverse clinical syndromes, ranging from relatively mild IgE-mediated allergy in allergic subjects with rhinitis or asthma to infection progressing to lethal disseminated tissue invasion in patients with severely compromised immune function. Within this spectrum lies allergic bronchopulmonary aspergillosis (ABPA), a unique type of allergic pulmonary disorder caused by a combination of innate and adaptive immune responses to *Aspergillus fumigatus* leading to severe pulmonary inflammation and tissue damage. ABPA is most common among patients with cystic fibrosis and asthma and is also seen in heterozygous carriers of cystic fibrosis mutations and in individuals with mutations affecting surfactant proteins.

The spores of *A. fumigatus* are 2–3 μm in diameter, and their small size allows them to pass through the narrowest airways and enter the alveoli when inhaled. Normally these spores are removed by mucociliary clearance, a mechanism whereby a thin fluid film of mucus covering the airway lining traps inhaled particulate matter and is constantly swept upwards towards the pharynx, where it is then either swallowed or expelled by coughing. This mechanism is disrupted in some settings. Patients with asthma may have overproduction of mucus and mucus plugging of the airways, and those with cystic fibrosis produce unusually viscous mucus by virtue of mutations i n the *CFTR* gene. Spores not cleared from the airway are viable and germinate, forming hyphae that release enzymes (superoxide dismutase, catalases, and proteases) and toxins that disrupt the epithelial barrier, further decrease mucociliary function, and activate innate immune responses in the lung.

The spores and hyphae of *A. fumigatus* are rich in molecules that can activate the innate immune system, and this may explain their tendency to induce inflammatory responses. The innate immune system relies on a finite number of receptors of committed specificity for molecular structures common to pathogens. These are encoded in germline DNA and have been evolutionarily conserved from before the separation of plants and animals. Although limited in repertoire compared with the amazing custom-tailored diversity that can be generated by adaptive immune responses, the innate immune system is preprogrammed and can immediately deal with challenges from pathogens. Innate immune responses provide a critical first line of defense against tissue penetration by fungi in the airways, and defects in innate immunity are

Topics bearing on this case:
Inflammation
Innate immune responses
T_H17 responses

This case was prepared by Hans Oettgen, MD, PhD, and Raif Geha, MD, in collaboration with Michael Young, MD.

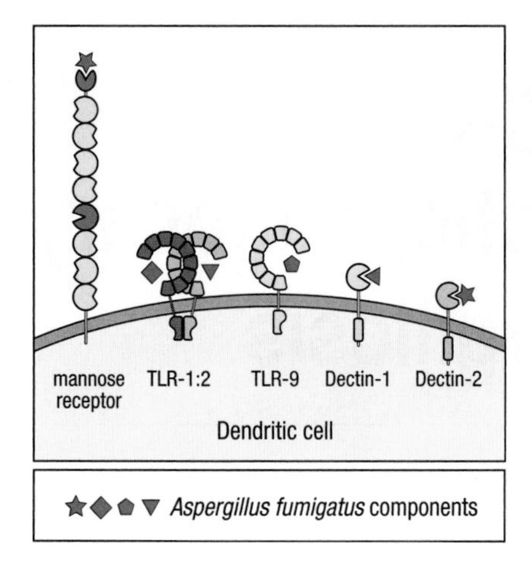

Fig. 14.1 Fungal components and their innate immune receptors. The cell membrane of fungi such as *Aspergillus* are surrounded by layers of chitin, β-glucans and mannoproteins. Innate immune effector cells, including dendritic cells residing in the airway, express several families of receptors that can recognize these fungal components and then activate both innate and adaptive immune responses. Mannose on mannoproteins (star) engages the mannose receptor. Both chitin (square) and β-glucans (triangle) can activate dendritic cells via Toll-like receptors (TLR-1:2). Fungal DNA can activate TLR-9. β-Glucans also signal via the Dectin-1 receptor, and mannoproteins via Dectin-2.

associated with invasive fungal disease. However, when confronted with a persistent stimulus or heavy pathogen load, the effector cells of innate immunity can drive a chronic state of inflammation, leading in some cases to tissue damage.

A. fumigatus spores and hyphae are laden with ligands recognized by innate immune receptors. Their cell walls contain chitin (β-(1-4)-poly-*N*-acetyl D-glucosamine), glucans (polyglucose; β-1,3-D-glucan) and mannose-containing proteins. Their DNA contains CpG motifs, sequences with unmethylated cytidines. Glucans are ligands for the innate immune receptors Dectin-1 and the Toll-like receptors TLR-1:2 and TLR-6. Chitin seems to signal via TLR-2 as well, and the mannoproteins bind to the mannose receptor. Fungal DNA is recognized by TLR-9 (Fig. 14.1). The interaction of each of these ligands with their innate immune receptors provides a danger signal for lung dendritic cells. This signal not only activates the dendritic cells to take on an antigen-presenting phenotype and migrate to regional lymphoid tissues, but also influences the polarity of the adaptive helper T-cell response they induce. Chitin-stimulated dendritic cells in particular have been shown to induce T_H17 responses, whereas dendritic cells activated by glucans via Dectin-1 promote T_H2 responses (Fig. 14.2). Tumor necrosis factor-α (TNF-α) produced

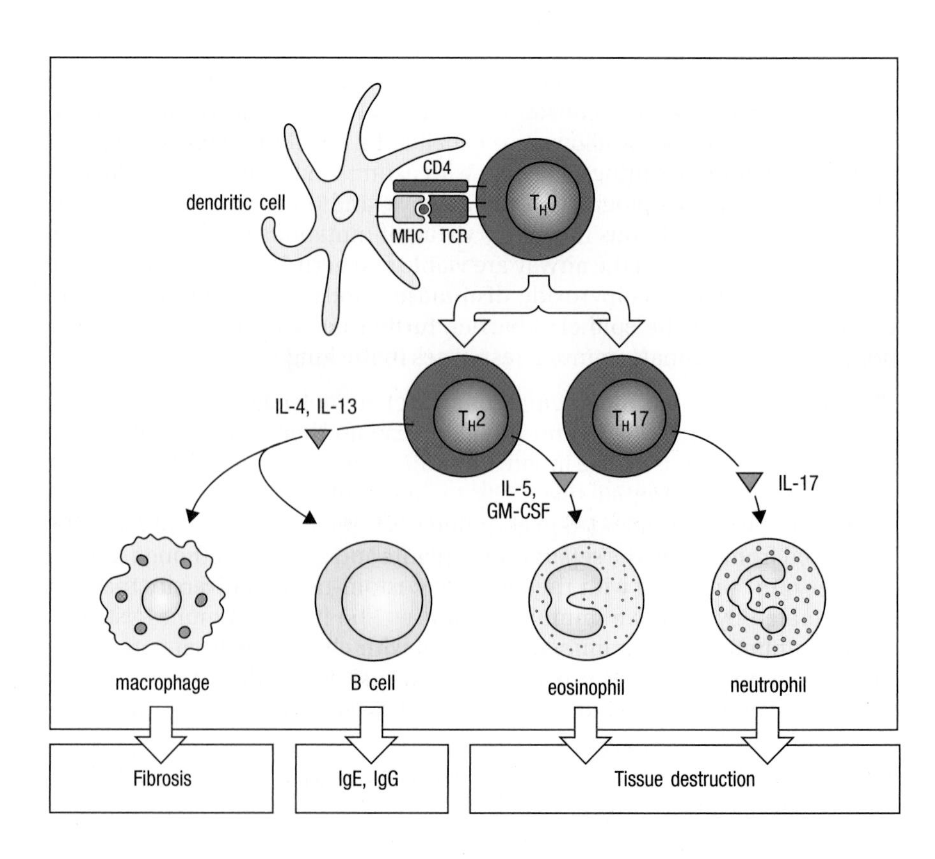

Fig. 14.2 Immunologic pathways driving tissue damage, fibrosis, and antibody production in allergic bronchopulmonary aspergillosis (ABPA). The key feature that distinguishes ABPA from allergic asthma is the potential to progress to tissue damage and fibrosis. The nature of the inflammatory response observed in the lungs of patients with ABPA suggests a mixed T_H2/T_H17 process. Persistent *A. fumigatus* mycelia in the airway (as can occur with mucociliary dysfunction in cystic fibrosis and perhaps in mucus overproduction in asthma) lead to epithelial damage and exposure to the innate effector cells of the airway, including dendritic cells. These in turn activate helper T (T_H) cells, which can differentiate into T_H2 and T_H17 cells. IL-4 and IL-13 produced by *A. fumigatus*-responsive T_H2 cells activate B cells to produce IgE and IgG antibodies. The same cytokines induce alternatively activated macrophages (AAM), which mediate tissue fibrosis. IL-5, along with granulocyte–macrophage colony-stimulating factor (GM-CSF) from T_H2 cells induces eosinophilopoiesis and eosinophil recruitment. T_H17 cells, which produce IL-17-family cytokines, are important in the recruitment of neutrophils in response to fungal antigens. Toxic products of eosinophils and neutrophils mediate tissue destruction.

by dendritic cells activated by innate immune receptor ligands enhances the expression of leukocyte adhesion molecules on the vascular endothelium of local blood vessels leading to local accumulation of inflammatory cells.

T_H2 cells are closely linked to allergy and are characterized by their production of cytokines that promote IgE production (IL-4 and IL-13) and eosinophilopoiesis (IL-5 and granulocyte–macrophage colony-stimulating factor (GM-CSF)). T_H2 cells arise under the influence of IL-4. In ABPA, T_H2 cytokines are important in promoting IgE and IgG isotype switching in B cells, leading to the *A. fumigatus*-specific antibody responses typical of the early phases of the disease (see Fig. 14.2). In addition, the induction of eosinophil formation in the bone marrow, along with the induction of the eosinophil-attracting chemokines CCL11 and CC24 in respiratory epithelial cells by T_H2 cytokines, leads to the recruitment of eosinophils into tissues. Persistence of eosinophils along with their tissue-damaging proteins, major basic protein and eosinophil cationic protein, contribute to local tissue damage and fibrosis. In chronic allergic responses such as ABPA, T_H2-derived IL-4 and IL-13 can lead to 'alternative' macrophage activation (see Fig. 14.2). Classically activated macrophages are stimulated by the T_H1 cytokine interferon-γ to activate the intracellular killing of intracellular pathogens. In alternative macrophage activation, IL-4 and IL-13 induce a distinct macrophage program, including the induction of arginase, an enzyme important in the biosynthesis of proline, a key component of collagen. Alternatively activated macrophages are likely to have a role in the late-phase fibrotic changes seen in ABPA.

In contrast to T_H2 cells, T_H17 cells are induced by the cytokines IL-6 and transforming growth factor (TGF)-β and express the transcription factor RORγT. They produce IL-17-family cytokines that stimulate the production of neutrophil-attracting chemokines by fibroblasts, macrophages, and epithelial cells, leading ultimately to neutrophil-mediated inflammation. Neutrophil-derived proteases and elastase as well as reactive oxygen species generated by activated neutrophils are all possible contributors to the pulmonary infiltrates, mucus plugging, microabscesses, exudative bronchiolitis, and central bronchiectasis observed in late-phase ABPA.

The case of Josephine March: a 14-year-old girl with worsening asthma.

Josephine was referred to the Allergy Program at Children's Hospital for evaluation of severe asthma. She had no history of pulmonary symptoms until she was 11 years old, when she developed a chronic cough with episodic wheezing. The family was living in Florida at the time, and the onset of her symptoms coincided with their house being flooded from Hurricane Wilma, with subsequent mold contamination of the basement and first floor. Despite mold remediation of the home, Josephine's cough and wheezing became progressively worse and she was admitted to hospital. A chest X-ray showed infiltrates and she was treated with antibiotics, nebulized bronchodilators, and systemic steroids, after which her symptoms improved. She was maintained on inhaled steroids, but when her symptoms failed to resolve she was diagnosed with moderate persistent asthma. Josephine's symptoms were triggered by mold, dust, and cat dander, but not by exercise. She was evaluated by a local allergist and had positive skin tests to tree pollens, dust mites, cat and dog dander, and molds. She was started on a combination of inhaled glucocorticoid and long-acting beta-agonist (fluticasone/salmeterol dry powder inhaler 100 μg/50 μg), using one puff twice a day to reduce airway inflammation and bronchospasm. Josephine was also given a short-acting beta agonist (albuterol inhaler) to use as needed, and

14-year-old girl with
mold exposure and
worsening asthma.

the long-acting histamine-receptor blocker loratidine, 10 mg once daily. On this regimen, her symptoms improved but did not resolve.

At age 13 years, Josephine moved to Boston, Massachusetts. She continued to have episodic wheezing and cough despite her maintenance medications. Exacerbations of her asthma led to her missing 10–14 days of school in that year, and short courses of oral prednisone were required three to four times in the year to control flares that were unresponsive to her baseline medications. She was referred to the Children's Hospital Allergy Clinic for further evaluation.

Josephine had no history of recurrent infection; her only infection was the pneumonia at age 11 years. Her new home in suburban Boston had no pets, cigarette smokers, or evidence of mold.

On physical examination, Josephine was a pleasant 14-year-old girl with mild darkening of the skin around the eyes ('allergic shiners'). Her nasal turbinates were edematous with clear secretions. Her lungs were clear. She had no digital clubbing.

Reversible small airway obstruction.

Measurement of airflow with a spirometer in the clinic showed significant obstruction. Her FEV_1 (the forced expiratory volume in 1 second) showed a 15% improvement after administration of two inhalations of the short-acting beta agonist albuterol, consistent with the type of reversible airflow obstruction that is seen in patients with asthma. Skin-prick tests for allergy were positive to dust mites, tree and weed pollens, cat and dog dander, and the molds *Aspergillus*, *Alternaria*, and *Penicillium*. A chest X-ray showed peribronchial cuffing, a sign of swelling of the lining of the airways and hyperinflation that indicates trapping of air in the lungs as a result of diminished expiratory air flow. Josephine's serum IgE was elevated at 812 IU ml^{-1} (normal less than 200). Her complete blood count was normal, but 10% eosinophils were noted on her differential count (normal 2–4%). Because of her clinical history of mold-exacerbated symptoms and suboptimal response to treatment, the positive skin test to *A. fumigatus*, elevated IgE, and eosinophilia, tests for *Aspergillus* precipitins (IgG antibodies that form immune complexes with *A. fumigatus* antigens, detected as precipitates in the laboratory) and *A. fumigatus*-specific IgE were ordered. These were positive, and the diagnosis of ABPA was made for Josephine.

Elevated IgE and peripheral blood eosinophils along with Af-specific IgG: diagnosis ABPA.

Josephine was started on the corticosteroid prednisone at 25 mg (0.5 mg per kg body weight) daily for 2 weeks and continued to take her fluticasone/salmeterol inhaler. Her chronic cough and wheezing soon improved, and the need for rescue albuterol inhalations diminished. By the third week of daily prednisone her symptoms resolved. Josephine's prednisone dose was tapered to 25 mg every other day and then continued for another 8 weeks. Her IgE was re-tested and had decreased to 300 IU ml^{-1}. The infiltrates previously observed on chest X-ray had cleared. In response to her clinical improvement and decreased IgE, prednisone was very gradually decreased over the next 3 months. When the dose reached 5 mg every other day, Josephine's IgE had fallen to 150 IU ml^{-1} and she remained asymptomatic. Her prednisone was discontinued.

Josephine did well for 4 months on her maintenance inhaled fluticasone/salmeterol and albuterol as needed. Unfortunately, the following spring, her cough and wheezing recurred. She was treated with prednisone with temporary improvement but subsequently needed two more courses of the drug. Her repeat IgE was elevated at 740 IU ml^{-1}, close to the level at the time of her ABPA diagnosis. Josephine did well on 25 mg prednisone daily, but her symptoms recurred when the dose was tapered below 20 mg. A repeat chest X-ray showed scattered bilateral infiltrates. Continued treatment with 20 mg prednisone daily and a 10-day course of amoxicillin–clavulonic acid (to treat a presumed bacterial pneumonia) failed to clear the infiltrates. A chest CT scan showed no bronchiectasis (a condition arising in some patients with ABPA in which there is irreversible destruction of the connective tissues and muscle of the large airways, leading to dilation and a tendency to airway collapse, resulting in

airflow obstruction). The antifungal agent itraconazole was added to her regimen, and over the course of 10 days her symptoms resolved; after a month a repeat chest X-ray showed that all infiltrates had cleared. She remained on 20 mg prednisone every other day for 4 months before her IgE decreased to 375 IU ml^{-1}, and the prednisone dose was successfully tapered over the next 8 months.

Allergic bronchopulmonary aspergillosis.

Allergic bronchopulmonary aspergillosis (ABPA) was first described in 1952 by Hinson and colleagues. From various surveys in the United States and worldwide it affects 1–2% of asthmatics, 7–14% of steroid-dependent asthmatics, 39% of asthmatics admitted to intensive care units, and 2–15% of patients with cystic fibrosis. The typical clinical features of ABPA are chronic persistent asthma, pulmonary infiltrates, and central bronchiectasis, but patients can exhibit a wide spectrum of symptoms, ranging from mild wheezing and cough to end-stage pulmonary fibrosis. The lung damage observed in this disease (bronchiectasis and fibrosis) is irreversible and distinguishes it from asthma. Patients experience episodic exacerbations characterized by wheezing, a productive cough with expectoration of brown–black mucus plugs, breathlessness (dyspnea) or hemoptysis (coughing up blood). Systemic symptoms such as low-grade fever, malaise, and weight loss can occur with exacerbations. Clubbing of the digits is reported in 16% of patients. Patients with ABPA can also be asymptomatic, only being diagnosed on screening as a result of chronic or worsening asthma.

ABPA should be considered in patients with chronic asthma or cystic fibrosis with a worsening course, increased sputum production, and/or chest X-ray abnormalities. The hallmark of ABPA is elevation of total serum IgE above 1000 IU ml^{-1}. The IgE level is affected by corticosteroid treatment, so a lower IgE level obtained in a patient already on systemic steroid treatment does not rule out ABPA. Disease severity can be monitored by performing serial measurements of total IgE levels. Levels of *A. fumigatus*-specific IgE and IgG and precipitating IgG antibodies are all elevated. A positive skin test for *A. fumigatus* is a major criterion for diagnosing ABPA. Peripheral blood eosinophilia at greater than 1000 cells μl^{-1} can also occur but is not essential for diagnosis. Chest radiographic findings are variable, including transient or fixed infiltrates, 'finger-in-glove' appearance as a result of mucus plugging and lung collapse. Chest CT imaging is the best technique for visualizing the changes in ABPA, the hallmark of which is central bronchiectasis with peripheral tapering of bronchi, which is considered a pathognomonic finding—that is, diagnostic of that disease (Fig. 14.3). The presence of *A. fumigatus* in sputum cultures, expectoration of brown–black mucus plugs and delayed skin reactivity to the fungus are considered supportive but not diagnostic features of ABPA. The major and minor criteria for the diagnosis of ABPA are listed in Fig. 14.4. In asthmatics without bronchiectasis (like Josephine), a diagnosis of ABPA requires these criteria: positive skin tests for *A. fumigatus*, elevation of total IgE, *A. fumigatus*-specific IgE or IgG, and the presence of *A. fumigatus*-precipitating antibodies in the serum. A diagnostic algorithm for ABPA has been proposed (Fig. 14.5).

In 1977, Rosenberg and Patterson classified five clinical stages of ABPA (Fig. 14.6). These stages are not meant to imply a sequential progression of the disease—rather, they score disease activity and responses to therapy. Stage 1 represents new, active disease; stage 2 describes disease remission, during which total IgE levels decline by 35–50%. This decline in IgE is sufficient to ascertain clinical remission; normalization of total IgE levels is not required and many

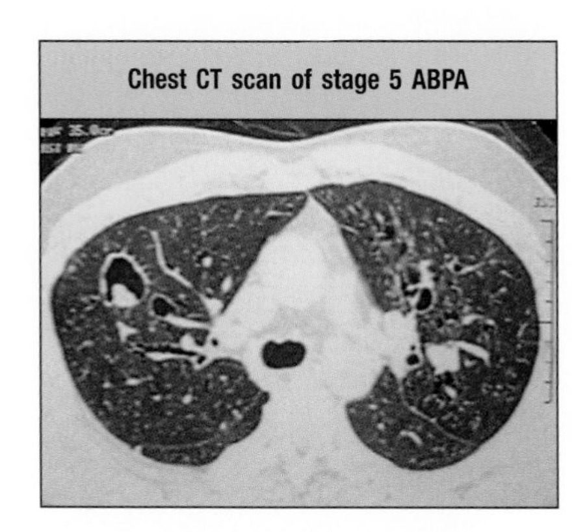
Chest CT scan of stage 5 ABPA

Fig. 14.3 Chest CT scan of stage 5 ABPA. This CT scan shows a cross section of the chest cavity. The patient is positioned with the back down. The dark spaces on each side are the lungs and they surround the central white area, the mediastinum. Inspection of the lungs reveals massively dilated central airways (bronchiectasis) and evidence of extensive fibrosis.

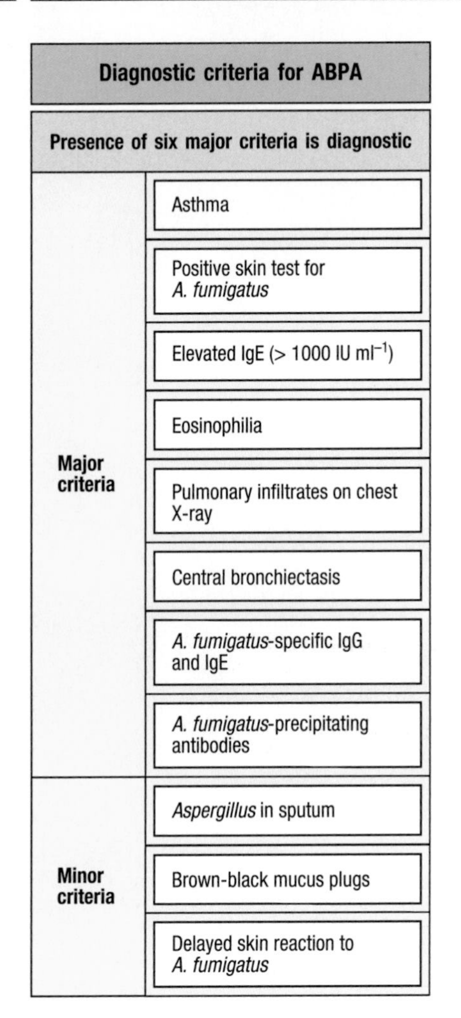

Diagnostic criteria for ABPA
Presence of six major criteria is diagnostic
Major criteria
Asthma
Positive skin test for A. fumigatus
Elevated IgE (> 1000 IU ml^{-1})
Eosinophilia
Pulmonary infiltrates on chest X-ray
Central bronchiectasis
A. fumigatus-specific IgG and IgE
A. fumigatus-precipitating antibodies
Minor criteria
Aspergillus in sputum
Brown-black mucus plugs
Delayed skin reaction to A. fumigatus

Fig. 14.4 Diagnostic criteria for ABPA.

patients with stage 2 ABPA continue to have persistently elevated IgE. Periodic IgE testing helps to establish the patient's baseline IgE level, which is useful for ongoing monitoring of disease activity. Treatment is typically required for 6–9 months. When there have been no exacerbations over a 3-month period, the ABPA is considered to be in remission. Serial IgE testing is then routinely performed every 6 months for a year, and then annually. Exacerbations can occur in 25–50% of patients after remission. Relapses (stage 3) are defined by a doubling of baseline total IgE. Patients who become steroid dependent, either because of their asthma or because of ABPA with elevated IgE levels, are in stage 4. Patients who develop fibrocavitary disease, fibrosis, and airway dilation (stage 5) may progress to respiratory failure and death. However, even in stage 5, patients can continue to be steroid responsive. There are no predictive indicators to identify patients at risk for progression to stage 5 disease. The rationale behind early consideration and diagnosis of the disorder in poorly controlled asthmatics and the subsequent institution of aggressive treatment is to prevent progression of the disease to stage 5.

Oral corticosteroids are the mainstay of treatment. As in Josephine's case, they are typically given at a dose of 0.5 mg prednisone per kg body weight for 2 weeks, followed by alternate-day therapy for 6–8 weeks. Subsequently, total IgE levels and chest X-ray findings are used to guide therapy. A decline in total IgE by 35–50% is a satisfactory response to treatment; doubling of IgE indicates a relapse. With remission, the prednisone dose is reduced by 5–10 mg every 2 weeks until it is stopped. If steroid taper is unsuccessful, a patient is in stage 4 and is treated with alternate-day steroids at the lowest possible dose.

Antifungal agents have been used to treat ABPA, but long-term controlled studies supporting their efficacy are lacking. Several short-term studies have demonstrated a reduction in total IgE levels by 25% or more, decreased sputum eosinophilia, and a reduction in oral steroid use. Itraconazole is the antifungal agent most commonly used to treat ABPA in current practice, but because of the limited evidence for its efficacy, and multiple adverse effects, including elevation of liver enzymes and inhibition of methylprednisolone (but not prednisolone) metabolism, its use should be considered on a case-by-case basis.

In various studies, between 2% and 15% of patients with cystic fibrosis developed ABPA. The complications of ABPA in cystic fibrosis are deteriorating lung function and clinical course, often with hemoptysis and pneumothorax (partial collapse of the lung resulting from an air leak into the pleura, the membranes surrounding the lung). The diagnosis of ABPA in cystic fibrosis is difficult, because cystic fibrosis itself manifests with the chronic obstructive respiratory symptoms, mucus plugging, bronchiectasis, and pulmonary infiltrates seen in ABPA. Atopy is a risk factor for ABPA in patients with cystic fibrosis: 22% of atopic patients develop ABPA, in contrast with 2% of non-atopic patients. In addition, there is a 34% frequency of sensitization in cystic fibrosis. The current consensus approach is to screen cystic fibrosis patients older than age 6 years for ABPA by annual total serum IgE determinations, with skin testing for *A. fumigatus* if the IgE level is greater than 500 IU ml^{-1}. If the skin test is positive, the diagnosis can be made on the basis of a history of acute clinical deterioration and the presence of *A. fumigatus*-specific IgG, IgG precipitins, or changes in chest X-ray or CT imaging studies. The treatment of ABPA in cystic fibrosis is the same as that recommended in other patients.

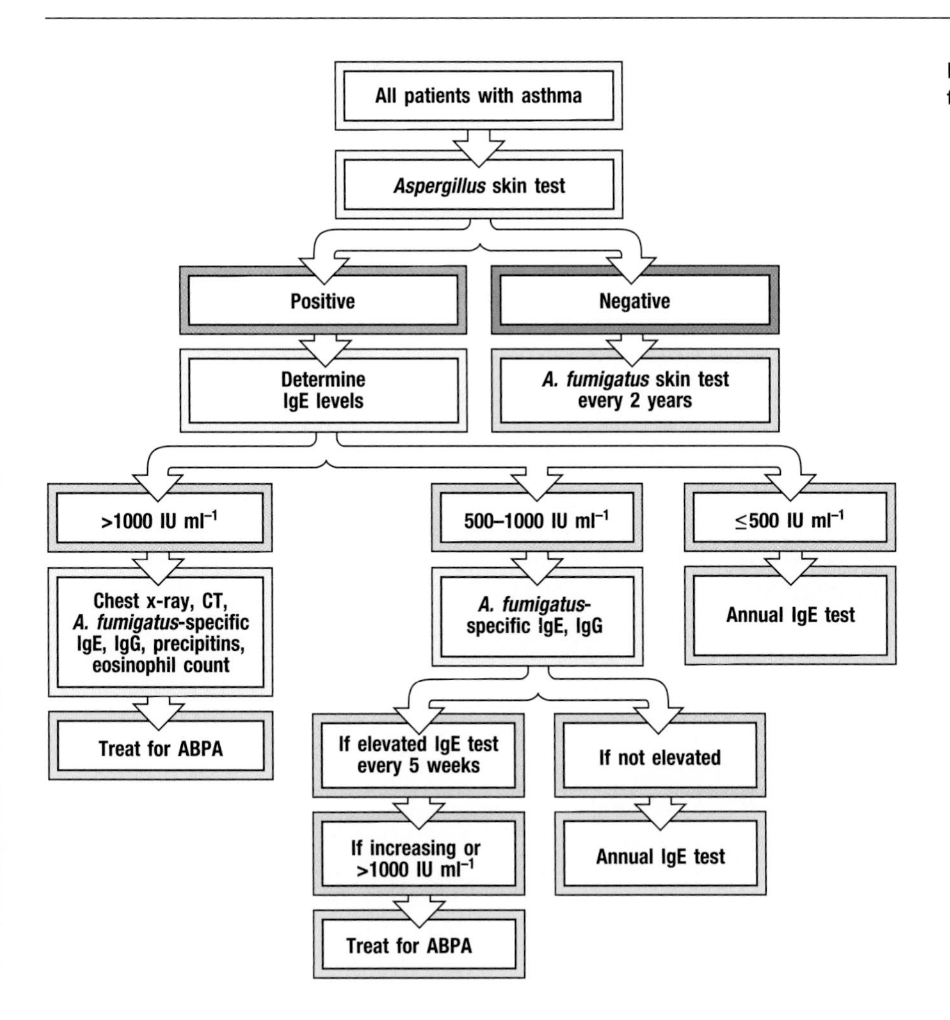

Fig. 14.5 Diagnostic algorithm for ABPA.

Clinical stages of ABPA				
Stage	**Description**	**Clinical features**	**Radiologic findings**	**Immunologic features**
1	Acute	Wheezing, fever, weight loss	Normal or opacities	High IgE, presence of *A. fumigatus*-specific IgE, IgG, precipitins
2	Remission	Asymptomatic	Normal or significant resolution from stage 1	35–50% decline in IgE by 6 weeks to 3 months
3	Exacerbation	Same symptoms as stage 1	Transient or fixed opacities	Doubling of IgE from baseline
4	Steroid-dependent	Symptomatic	Transient or fixed opacities	**Group 1** Non-rising IgE, steroids needed for asthma control **Group 2** Steroids required to suppress high IgE
5	End stage fibrotic	Symptomatic, severe, fixed obstruction, cor pulmonale	Bronchiectasis, pulmonary fibrosis, pulmonary hypertension	Elevated IgE, specific IgE, IgG

Fig. 14.6 Clinical stages of ABPA.

Questions.

1 *What immune mechanisms normally clear A. fumigatus from the lungs?*

2 *Do you think that ABPA could develop in non-atopic individuals?*

3 *Compare and contrast the immunologic mechanisms in ABPA with A. fumigatus hypersensitivity that causes allergic rhinitis or asthma.*

4 *Josephine did not have bronchiectasis. What criteria led to her diagnosis of ABPA?*

5 *What was Josephine's clinical stage when she was treated with itraconazole?*

CASE 15 Hypersensitivity Pneumonitis

A non-IgE-mediated hypersensitivity response to an aeroallergen.

The lungs and airways are constantly exposed to microscopic inhaled airborne particles floating in the air. Under normal conditions, the protein antigens carried on these particles neither elicit measureable immune responses nor induce hypersensitivity reactions. In atopic individuals, IgE responses to airborne allergens can trigger the activation of mast cells and basophils in type I immediate hypersensitivity reactions manifesting as asthma (see Case 2) or allergic rhinitis (see Case 3), afflictions affecting up to 10–30% of the population. Aeroallergens include pollens, molds, and animal proteins, all of which have a propensity to induce T_H2-cell responses. T_H2 cells provide cytokines important for IgE isotype switching (interleukin (IL)-4 and IL-13) as well as for eosinophil development and recruitment (granulocyte–macrophage colony-stimulating factor (GM-CSF) and IL-5). Both the induction of IgE production and the elicitation of symptoms can be triggered by minute amounts of inhaled allergen.

A much smaller proportion of individuals develop a very distinct antigen-driven lung disease that can be triggered by the inhalation of large quantities of a wide array of organic and inorganic antigens and has variously been labeled hypersensitivity pneumonitis, allergic alveolitis, or hypersensitivity lung disease. Acute symptoms typically develop within hours of heavy exposure to allergens and consist of dyspnea, cough, and fever. With low-grade recurrent or chronic exposure, a more insidious pattern of chronic or persistent dyspnea, cough, anorexia, and fatigue often emerges. The pathology of this condition affects the alveoli and terminal airways. In contrast to the allergic airway diseases, it seems to arise by IgE-independent mechanisms and is not typically associated with eosinophil infiltration into the affected tissues. Hypersensitivity pneumonia is also clinically distinct from allergic disorders in that it tends to elicit systemic symptoms, including fever and fatigue, and is associated with irreversible airway changes including fibrosis. High-titer antigen-specific IgG antibodies—precipitins—are present in 75% or more of individuals with this condition, and bronchoalveolar lavage fluid analysis by flow cytometry reveals an expansion of CD8 T cells. The affected airways exhibit a variety of granulomatous and fibrotic changes. Taken together, the immunological and pathological features of hypersensitivity pneumonitis suggest mechanisms of pathogenesis that combine type II, III, and IV mechanisms (see Fig. 4.1).

Topics bearing on this case:

Types of hypersensitivity reactions

Allergic responses to aeroallergens

This case was prepared by Hans Oettgen, MD, PhD, and Raif Geha, MD, in collaboration with Frank J. Twarog, MD, PhD.

Frank Wheeler: a 39-year-old man with breathlessness and fever.

39-year-old man with fever, dyspnea, and cough.

Frank Wheeler, a successful 39-year-old corporate attorney, attended the Allergy Clinic for evaluation of chronic cough, dyspnea (shortness of breath), and fever. He had been in general good health, was on no routine medications, and was a lifelong nonsmoker. He was an avid athlete, playing competitive tennis at his club. Around 6–7 months previously, Frank began to notice some exertional dyspnea while playing tennis. The symptoms were fairly mild and varied in severity. He reported the symptoms to his primary care physician. On examination, there were no obvious cardiac or pulmonary findings. His chest was clear, without crackles, wheezes, or reactive cough on expiration.

Because Frank had a history of mild seasonal rhinitis and a positive family history of reactive airway disease, the possibility of exercise-induced bronchoconstriction was raised. A short-acting beta-adrenergic agent (albuterol) was prescribed, to be taken by inhalation 15–30 minutes before vigorous exercise. Over the next few weeks, he did not notice any significant benefit when using the bronchodilator and he returned to his primary care physician. Once again, his chest examination was normal. A stress test, performed to rule out cardiac causes of exertional dyspnea, was normal. Because of a family history of chronic obstructive pulmonary disease, the possibility of α_1-antitrypsin deficiency was considered and an α1-antitrypsin level and Pi type (analysis of α_1-antitrypsin isoforms) were obtained. The level was within the normal range and Pi type M (normal).

Recurrent flu-like symptoms, crackles on chest exam.

Frank frequently traveled in the course of his work. On one occasion, within 12–24 hours of returning from a European business trip, he developed a moderately severe, non-productive cough, low-grade fever (temperature around 37.8°C), and mild, shaking chills. On examination, he had scattered inspiratory crackles. He was clinically diagnosed with either viral or mycoplasmal pneumonia and the antibiotic azithromycin was prescribed. He experienced several episodes of flu-like symptoms over the next month or two. These were treated symptomatically with nonsteroidal anti-inflammatory agents. Frank's respiratory symptoms flared up again after another extended business trip. He noticed that he never had the symptoms while traveling and expressed concern that sensitivity to mold or some other environmental factor in his house might have been responsible for his symptoms. An allergy consultation was recommended.

Physical examination in the Allergy Program was generally unrevealing. There was no conjunctival redness, edema of the nasal mucosa, or nasal polyps. On chest auscultation, a few crackles were heard. Pulmonary function studies indicated a restrictive pattern, with a decreased forced vital capacity (FVC) at 60% of normal but a normal ratio of the forced expiratory volume in 1 second (FEV_1) to FVC, consistent with normal airflow (indicating no obstruction such as is seen in asthma) (Fig. 15.1). Frank's

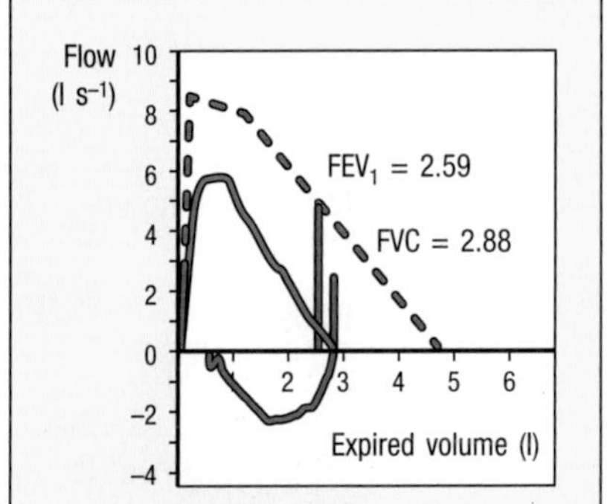

Fig. 15.1 Lung function as assessed by spirometry. Airflow is measured during a forced expiration and plotted as flow (liters s^{-1}) against expired volume (liters) in the green tracing above the *x*-axis. The subsequent inspiration is indicated by the blue tracing below the *x*-axis. The predicted expiratory flow tracing for a normal subject of the same height is indicated by the broken line. The trace shown here reveals a decreased forced vital capacity (FVC), indicating that the patient exhaled a smaller total volume of air than expected. Vertical markers indicate the volume expired in the first second (FEV_1) and the total expired volume (FVC). The FEV_1 is reduced to the same extent, so that the ratio, FEV_1/FVC, a measure of airflow, is normal, suggesting the absence of obstruction. This result is consistent with a restrictive, not an obstructive, process.

blood O_2 saturation was normal at 95%. Skin-prick testing and blood tests were performed to screen for allergen-specific IgE antibodies. Consistent with Frank's history of seasonal rhinitis, positive tests were noted for several tree and grass pollens. Test results for a wide variety of molds, done because of his concerns regarding household allergens, were, however, negative. Screening for IgG-mediated sensitivity to environmental antigens was performed by ordering a 'hypersensitivity precipitin panel.' This showed no evidence for IgG antibodies against *Aspergillus fumigatus*, *Saccharopolyspora rectivirgula* (*Micropolyspora faeni*), thermophilic actinomycetes, or pigeon allergens. A complete blood count was normal and showed no increase in circulating eosinophils. Total IgE was 100 IU ml^{-1} (minimally elevated). Immunoglobulin levels were measured because of the history of pneumonia. Frank's IgG was slightly elevated at 2200 mg dl^{-1}. A mannitol inhalation challenge was performed to evaluate airway reactivity and was normal. The sensitivity of mannitol challenge is excellent in detecting exercise-induced asthma, and the negative result argued against asthma as a cause of Frank's exertional dyspnea.

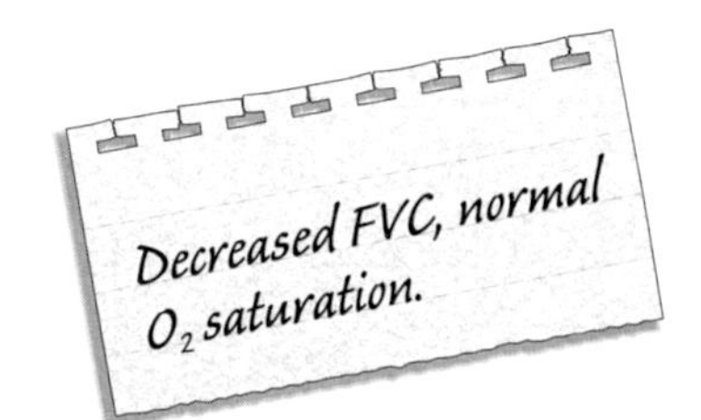

Frank's allergist considered other possible environmental causes for the cough. As an attorney, Frank was not exposed to any unusual occupational chemicals or allergens that could account for his symptoms. He had built a new home some 6 or 7 years earlier. It had hardwood floors with area carpets, the basement was dry, there were no household pets, and an air conditioner was routinely used during the warm months, but this was serviced regularly with replacement of filters. Of note, however, was the fact that a hot tub had been installed inside the home when it was built, and Frank used the tub frequently, especially when he returned home from stressful business trips. Analysis of the hot-tub water detected *Mycobacterium avium-intracellulare* complex, a combination of mycobacteria (*M. avium* and *M. intracellulare*) occasionally present in the warm water of hot tubs. Given Frank's history, especially the correlation between the time when he used the hot tub and the occurrence of his cough, along with the findings on chest examination and pulmonary function studies, the possibility of hypersensitivity lung disease emerged. A chest X-ray revealed patchy nodular infiltrates. A high-resolution chest CT scan was performed. Diffuse, patchy, ground-glass opacities and some central nodules were observed (Fig. 15.2).

M. avium-
intracellulare
complex in the hot
tub. Hot-tub lung?

To confirm the diagnosis, a bronchoscopy was performed. Studies on bronchoalveolar lavage (BAL) fluid, obtained by instilling a saline solution into the lungs via the bronchoscope and then recovering the fluid, showed a predominance of lymphocytes, which were largely CD8 T cells. *Mycobacterium avium* complex was isolated from respiratory secretions. In view of these findings, it was not felt that an open-lung biopsy was necessary. A diagnosis of hypersensitivity lung disease, 'hot-tub lung,' was made. Because Frank was not acutely ill at the time of his evaluation, systemic steroid therapy was deferred. He stopped using the hot tub until it had been thoroughly cleaned, and he had no further episodes of dyspnea and/or cough. At a follow-up with the allergist a year later, Frank was symptom-free and his exercise tolerance had returned to normal. His pulmonary functions had normalized, with FVC now at 98%.

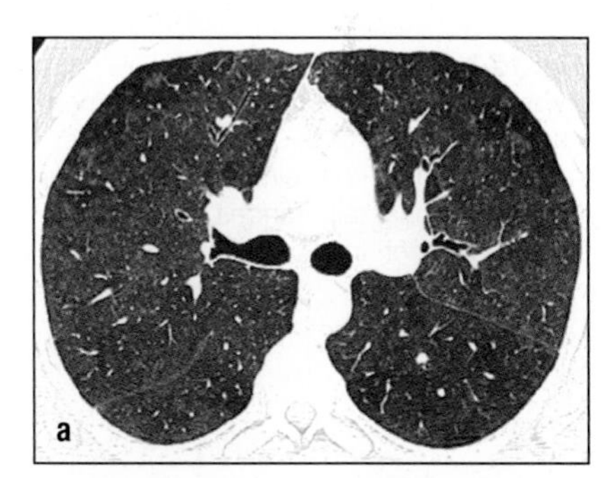

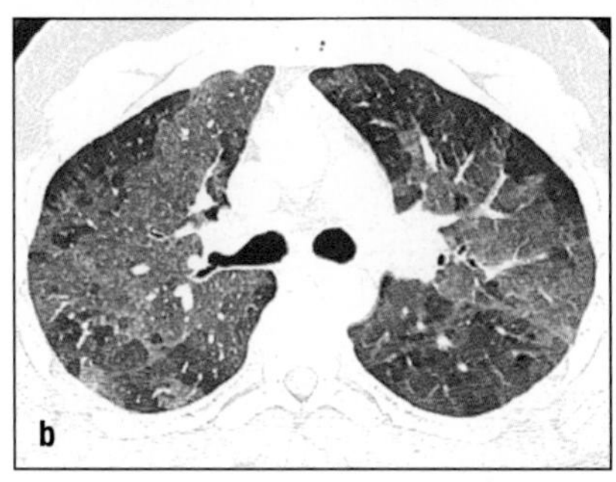

Fig. 15.2 High-resolution computed tomography of the chest. The scan reveals patchy areas of ground-glass attenuation and ill-defined centrilobular nodules scattered in both lungs (panel a) with air trapping on expiration (panel b).

Hypersensitivity pneumonitis.

Hypersensitivity pneumonitis, otherwise known as allergic alveolitis or hypersensitivity lung disease, is an immunologically mediated lung condition that involves the alveoli and terminal airways. In susceptible individuals it can be caused by inhalation of a wide variety of agents, including bacterial, fungal, and animal proteins as well as chemical agents such as isocyanates (Fig. 15.3). Curiously, fewer than 1% of individuals chronically exposed to these substances become symptomatic, although sensitization may occur in a larger percentage.

Agents most commonly associated with hypersensitivity pneumonitis		
Agent	**Source**	**Disease**
Microbes		
Thermophilic actinomycetes	Moldy plant materials	Farmer's lung
Saccharopolyspora rectivirgula (Micropolyspora faeni)	Moldy hay	Farmer's lung
Trichosporon cutaneum	Mold in Japanese homes	Summer-type hypersensitivity pneumonitis
Mixed amebae, fungi, and bacteria	Cold mist and other humidifiers, air conditioners	Nylon plant or office worker's or air pneumonitis air conditioner's lung, ventilation
Bacteria and fungi	Contaminated metal-working fluids	Machine-operator's lung
Animals		
Avian proteins	Bird excreta, blood, or feathers	Bird-breeder's lung, bird-fancier's lung, pigeon-breeder's lung
Chemicals		
Isocyanates	Paints, plastics	Paint-refinisher's lung
Anhydrides	Plastics	Chemical-worker's lung, plastic-worker's lung, epoxy-worker's lung

Fig. 15.3 Causative agents of hypersensitivity pneumonitis. Common causative agents of this disease include microbes, animal proteins, and chemicals. Many additional substances, not listed here, have been implicated as well.

Symptoms are varied but typically include dyspnea and cough in particular. Confusion surrounds the classification of hypersensitivity pneumonitis because there are no specific accepted criteria to distinguish between the various stages of this condition. In general, three clinical presentations are recognized: acute, subacute, and chronic. In the acute form, intermittent heavy exposure to the inciting agent results within hours in episodes of cough, dyspnea, and fever, often misleading the clinician into considering an infectious etiology. Symptoms usually remit spontaneously in 24–48 hours. Inspiratory crackles are heard on chest exam, and a chest X-ray may reveal patchy infiltrates. A subacute form is more subtle and insidious, with recurrent symptoms persisting for days to months. It is characterized by onset of cough, dyspnea, and fatigue several days to weeks after modest, persistent allergen exposure. Symptoms may occur episodically or evolve into a more persistent pattern. Interestingly, acute or subacute forms may persist even after prolonged allergen avoidance. Studies in Midwestern farmers with farmer's lung disease, a type of hypersensitivity pneumonitis, have shown that symptoms of the condition can recur in affected individuals returning to their farms after years of absence. In the chronic form, symptoms may be present for months to years. Dyspnea, persistent cough, weight loss, and anorexia are characteristic. The chronic form is characterized by diffuse fibrosis, and irreversible changes persist even when exposure to the inciting agent is limited. The response to systemic steroids is disappointing. This form is often difficult to distinguish from other causes of interstitial lung fibrosis, a group of lung diseases in which the interstitium but not the air spaces are affected by infection, connective tissue disorders, malignancy, or other inflammatory stimuli.

Because of the varying clinical presentations, the recognition of hypersensitivity pneumonitis may be diagnostically challenging. Several features of the clinical presentation are predictive of an ultimate diagnosis of hypersensitivity pneumonitis (Fig. 15.4). Most commonly, symptoms include dyspnea (more than 90% of cases) and cough (more than 60% of cases). Other presentations include recurrent flu-like symptoms, chest discomfort, and weight loss.

Because causative agents are often associated with specific occupations, hobbies, or environmental exposure, it has been common to label those specific conditions with the occupation or agent associated with the allergen exposure (see Fig. 15.3). For example, in the Midwest, dairy farmers may develop hypersensitivity pneumonitis from thermophilic actinomycetes in moldy hay: this condition has been called 'farmer's lung.' Exposure to avian proteins results in 'pigeon breeder's disease' in some individuals keeping pigeons; in England, budgerigars are associated with 'bird-fancier's lung.' In Frank's case, water contaminated with *M. avium-intracellulare* caused 'hot-tub lung.'

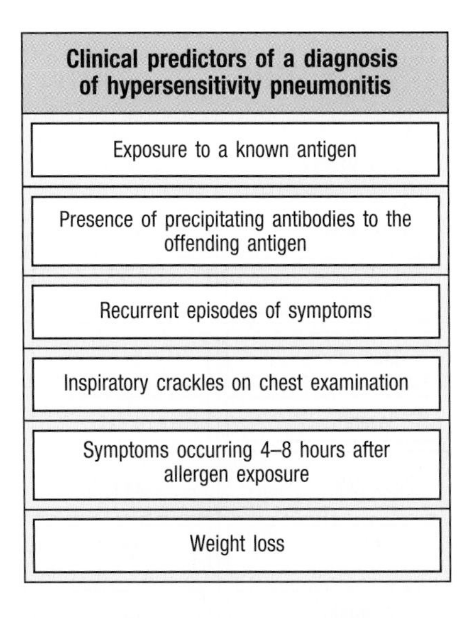

Clinical predictors of a diagnosis of hypersensitivity pneumonitis
Exposure to a known antigen
Presence of precipitating antibodies to the offending antigen
Recurrent episodes of symptoms
Inspiratory crackles on chest examination
Symptoms occurring 4–8 hours after allergen exposure
Weight loss

Fig. 15.4 Clinical predictors of a diagnosis of hypersensitivity pneumonitis.

A recent series reported from the Mayo Clinic studied 85 consecutive patients with hypersensitivity lung disease. The mean age of presentation was 53 ± 14 years, and 60% of patients were women. Only two patients were current smokers. The most common causes in the 64 patients with identified triggers were avian antigens (34%) and *M. avium* complex in hot-tub water (21%). Farmer's lung was diagnosed in 11% and heavy mold exposure in the home in 9%.

Questions remain about why only a small percentage of exposed individuals develop clinical symptoms, why acute episodes occur after prolonged periods of previous sensitization, and what leads to disease progression. The immunopathology of hypersensitivity lung disease also remains uncertain. Analysis of BAL fluid reveals that lymphocyte numbers are increased more than 25-fold and CD8 T cells are predominant. The CD4/CD8 T-cell ratio is decreased, and there is a higher proportion of γ:δ T cells. In the subacute forms of hypersensitivity lung disease, granulomas may develop, and lymphoid follicles containing plasma cells are noted. In the chronic fibrotic phase, alveolar macrophages have been found to express increased amounts of TGF-β, which is a potent stimulator of fibrosis and angiogenesis. Mast cells are sometimes recovered from BAL fluid and have a protease expression profile consistent with that of mast cells residing in connective tissue rather than mucosal tissues. This type of mast cell has been related to fibrosis. The combination of T cells and macrophages in BAL fluid and tissue fibrosis suggests that, despite its association with high-titer antibodies against the offending antigen, the dominant mechanism of hypersensitivity pneumonitis pathogenesis is likely to be a delayed-type hypersensitivity response (type IV reaction) (see Case 6). Concurrent viral or bacterial infections are believed to contribute to acute exacerbations in some cases. Patients occasionally report that, although they had been exposed to the suspected agent for some time, symptoms appeared after a recent acute respiratory infection.

Frank typifies some of the difficulties encountered in making a diagnosis of hypersensitivity pneumonitis. The initial presentation may be subtle and nonspecific. Other conditions, including viral infections or pneumonia, may be considered in the early phases of hypersensitivity lung disease or with recurrent, acute forms. Only after a more intense investigation of the environment and pattern of symptoms may the possibility of hypersensitivity lung disease become apparent.

Lung-function tests can provide important clues to this diagnosis. Two general patterns of abnormality can be detected by such testing: restrictive and obstructive. In restrictive lung physiology, the rates of airflow through the airways are normal but lung volumes are decreased. This can happen in the setting of fibrosis such as occurs in hypersensitivity pneumonitis. In contrast, obstructive physiology is characterized by decreased airflow with normal (or increased, because of air trapping) volumes. The obstructive pattern

Lung-function test results at presentation in hypersensitivity pneumonitis	
Type of abnormality	No. (%) of patients (n = 83)*
Obstruction	13 (16)
Mild	4
Moderate	5
Severe	4
Restriction	44 (53)
Mild	23
Moderate	10
Severe	11
Nonspecific abnormality	10 (12)
Isolated reduction in diffusing capacity	8 (10)
Normal	8 (10)

Fig. 15.5 Lung-function test results at presentation in hypersensitivity pneumonitis. *For two patients, lung-function data from the time of presentation were not available.

is characteristic of asthma (see Case 2). Pulmonary function tests in the 85 patients from the Mayo Clinic study are illustrated in Fig. 15.5, and suggest a restrictive pattern in many individuals. Obstructive findings are infrequent.

Routine laboratory or radiologic findings are often nonspecific. There is no eosinophilia on a complete blood count. During the acute phase, leukocytosis may be present. There may also be an increase in markers of acute inflammation, including the erythrocyte sedimentation rate and C-reactive protein. IgG may be modestly elevated (as it was in Frank's case), particularly with chronic or subacute disease. Positive testing for hypersensitivity lung disease precipitins (high-titer IgG antibodies against antigens associated with the disease) are helpful if positive, but their absence does not exclude the diagnosis. In a careful analysis of patients for a clinical diagnosis of hypersensitivity pneumonitis, only 78% had positive precipitins.

In patients thought to have hypersensitivity pneumonia, the diagnosis can be supported by BAL, lung biopsy, and radiographic studies. As noted above, BAL analysis reveals a lymphocytosis with a predominance of CD8 lymphocytes. The histopathology of hypersensitivity pneumonitis is varied. Biopsy findings may include peribronchial and interstitial lymphocytic infiltration, foamy macrophages, non-caseating granulomas, and fibrosis (Fig. 15.6). Routine chest X-rays are often normal, but high-resolution CT may reveal ground-glass opacities, centrilobular nodules, or a mosaic pattern. Other findings include fibrotic changes and peripheral honeycombing (see Fig. 15.2). In summary, a diagnosis of hypersensitivity pneumonitis usually evolves after a strong clinical suspicion arises from a careful medical history and is confirmed by an array of pulmonary-function studies, clinical imaging, laboratory, and pathology findings. The treatment is limited to antigen avoidance. Systemic steroids are occasionally used during acute exacerbations, but controlled studies of the response to this therapy are fairly limited.

Transbronchial biopsy in hypersensitivity pneumonitis

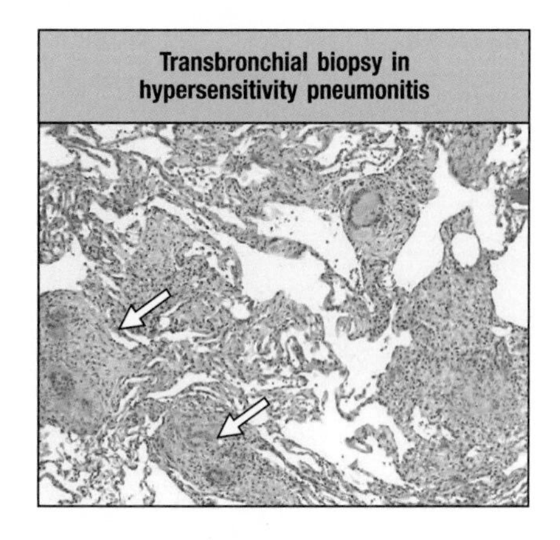

Fig. 15.6 Transbronchial biopsy in hypersensitivity pneumonitis. The photomicrograph shows a transbronchial biopsy taken from the right lower lung. It reveals a mononuclear cellular infiltration of the tissue spaces between the alveoli (chronic interstitial pneumonia) as well as several well-formed non-necrotizing granulomas (arrows). Photomicrograph: ×100; hematoxylin and eosin staining.

Questions.

1 *Of the three clinically described forms of hypersensitivity pneumonitis, which one did Frank display? Give the reasons for your answer.*

2 *Why would you perform an evaluation to rule out a diagnosis of asthma when presented with a patient with symptoms of hypersensitivity pneumonia that include breathlessness?*

3 *What specific therapies are used for the treatment of hypersensitivity pneumonitis? Why do you think treatment with corticosteroids is ineffective in the chronic form, even though it is a common treatment in other types of allergic inflammatory disease, such as asthma?*

4 *What typical findings are present in bronchoalveolar lavage in patients with hypersensitivity pneumonitis? What do they tell you about the underlying hypersensitivity mechanisms in this disease?*

CASE 16 Venom Hypersensitivity

A severe immunological reaction to an insect sting.

In Case 1 we saw how an allergen in peanuts can cause immediate life-threatening systemic anaphylaxis when eaten by a sensitized individual. That allergen was delivered into the bloodstream via the mucosa of the gastrointestinal tract and then distributed to mast cells residing in the cardiovascular and respiratory systems, leading to severe hypotension and difficulty in breathing. Both allergic sensitization (the induction of allergen-specific IgE antibody responses) and the elicitation of anaphylaxis can also be induced via other routes of allergen exposure, including inhalation, skin contact, and direct injection into the blood stream. The last of these routes, direct injection, occurs with the intravenous administration of medication and with insect stings, and has the capacity to introduce significant amounts of intact protein antigens directly into the circulation. These antigens are very rapidly transported directly to mast cells residing in the target organs of systemic anaphylaxis (Fig. 16.1). In sensitized individuals, IgE bound to these mast cells through its high-affinity receptor, FcεRI, then triggers immediate mast-cell activation with the release of preformed vasoactive mediators, including histamine, as well as the rapid synthesis of arachidonic acid metabolites (prostaglandins and leukotrienes). Symptoms induced by exposure to injected allergen in individuals with allergen-specific IgE antibodies vary depending on the magnitude of the IgE response and the number, tissue distribution, and sensitivity of mast cells. When more than one organ system is involved in the IgE-mediated reaction, it is termed systemic anaphylaxis. Anaphylaxis involves the skin and/or mucosa (generalized urticaria, flushing, and angioedema), the gastrointestinal tract (nausea, vomiting, diarrhea, and abdominal pain), the respiratory system (dyspnea, wheezing, stridor, and hoarseness), and the cardiovascular system (reduced blood pressure and circulatory collapse).

Accurate data on the prevalence of insect-sting reactions are hard to come by, and estimates vary depending on study methodologies. Hypersensitivity to insect venoms causes up to 30% of all anaphylactic episodes and about 20% of all fatal anaphylactic events worldwide. The stinging insects responsible for most allergic reactions belong to certain families within the order Hymenoptera: Apidae (bees), Vespidae (yellowjackets (wasps), hornets, and paper wasps), and Formicidae (ants) (Fig. 16.2).

Transient local reactions to insect stings are common and are characterized by a painful, erythematous swelling confined to the sting site(s) that subsides in a few hours. These reactions are probably late-phase IgE-mediated responses. In around 10% of affected individuals, the swelling is greater than 10 cm in

Topics bearing on this case:
Class I hypersensitivity reactions
Systemic anaphylaxis
Mast-cell activation via IgE
Allergen immunotherapy

This case was prepared by Hans Oettgen, MD, PhD, and Raif Geha, MD, in collaboration with Mona Hedayat, MD and Janet Chou, MD.

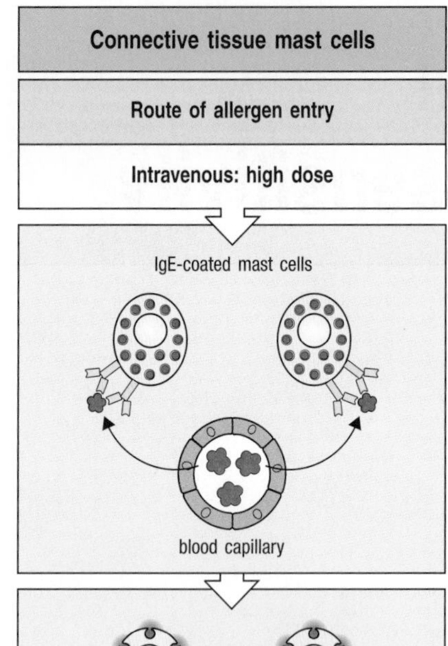

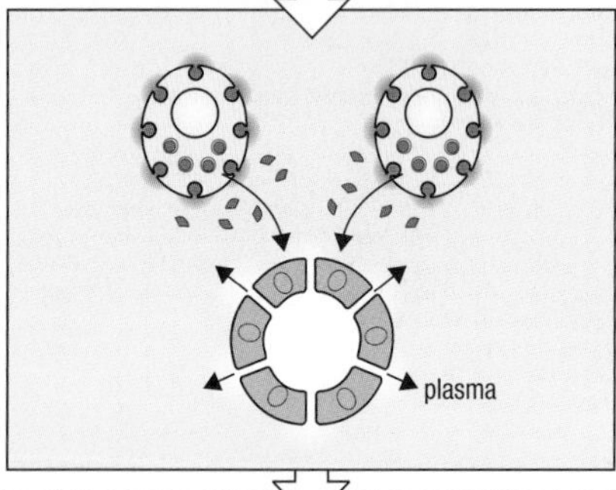

Fig. 16.1 Allergen entering the bloodstream directly can cause a systemic anaphylactic reaction. Allergen carried in the bloodstream (intravenous) activates mast cells throughout the body, resulting in the systemic release of histamine and other mediators, leading to anaphylactic shock.

diameter and is classified as a 'large local reaction.' The swelling can last up to 10 days. Estimates vary, depending on methods of data collection, but anaphylaxis occurs in approximately 0.3–7.5% of all stings.

In this case, we look at an anaphylactic reaction to multiple hornet stings in a child.

The case of Alec D'Urberville: an 11-year-old boy with an anaphylactic reaction to hornet stings.

On a hot summer day, Alec was climbing a tree when his foot slipped and he grabbed a branch to save himself from falling. A swarm of hornets flew out of a nest, stinging his face, arms, and legs many times. Within minutes, Alec developed large erythematous, swollen nodules at the site of each sting, followed by hives that spread all over his body. He began feeling tightness in his chest and difficulty in breathing. He did not show any angioedema, vomiting, or other symptoms. Fortunately, Alec's mother was nearby, and because Alec had an allergy to peanuts she always carried an epinephrine (adrenaline) injector for him. She made Alec lie down and injected the epinephrine into the side of his thigh while his father called 911. After a few minutes Alec's symptoms improved markedly.

When the emergency medical services arrived, Alec's heart rate was elevated but his blood pressure, respiratory rate, and oxygen saturation were all normal. He was given the antihistamine Benadryl (diphenhydramine), and taken to the hospital. By the time Alec arrived at the emergency room, his respiratory symptoms had resolved completely and the hives were considerably improved. A blood sample was sent for measurement of the enzyme tryptase. Alec was monitored overnight in the hospital and discharged the next day with recommendations to follow up with his allergist for testing.

11-year-old with hives and wheezing after a hornet sting: anaphylaxis!

At Alec's allergy clinic appointment 3 days later, his mother was disappointed to hear that a skin test could not be done immediately because it was too soon after the anaphylactic episode. The tryptase test taken in the emergency room revealed a mature tryptase level of 14 ng ml^{-1} (higher than normal). Alec's allergist reviewed the use of the EpiPen Jr epinephrine injector with his mother and scheduled a follow-up appointment 6 weeks later for diagnostic testing.

Family	Subfamily	Common name
Apidae		Honeybee, bumblebee
Vespidae	Vespinae	Yellowjacket, yellow hornet, white-faced hornet
	Polistinae	Paper wasp
Formicidae		Fire ant, jack jumper ant, harvester ant

Fig. 16.2 The families of hymenopteran insects whose stings can cause a systemic reaction.

Six weeks later, Alec had skin tests for honeybee, wasp, yellowjacket, and hornet venoms. Skin prick and intradermal testing was positive for hornet and yellowjacket venom. Because skin tests for reaction to honeybee and wasp venoms were negative, Alec was tested for IgE antibodies against these venoms. These tests were also negative. Baseline tryptase levels were ascertained during this visit. Alec had a total tryptase of 2 ng ml^{-1} and a mature tryptase level of less than 1 ng ml^{-1}, thus excluding a diagnosis of mastocytosis. Because of Alec's history of anaphylaxis to an insect sting as well as testing positive in skin tests, his allergist recommended starting venom immunotherapy ('allergy shots') for hornet and wasp venoms, with a long-term plan to continue the immunotherapy for at least 5 years.

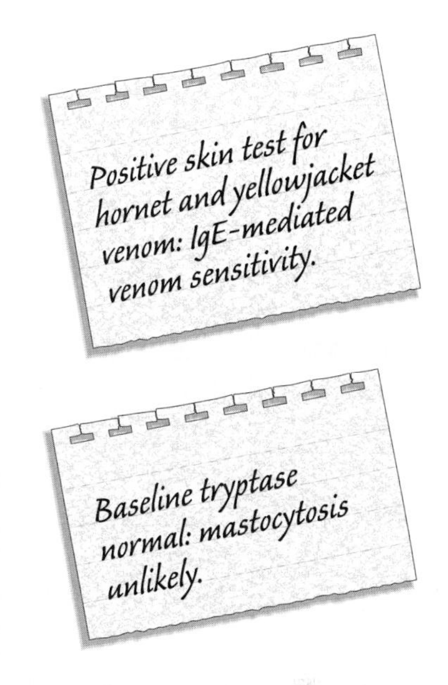

Positive skin test for hornet and yellowjacket venom: IgE-mediated venom sensitivity.
Baseline tryptase normal: mastocytosis unlikely.

Venom hypersensitivity.

Hypersensitivity to the venom of an insect is due to both IgE-mediated and non-IgE-mediated reactions (Fig. 16.3). The nonimmunologic effects of the venom can also result in toxic reactions and, rarely, multiorgan dysfunction. Allergic reactions are directed against the allergenic proteins of the hymenopteran venoms. There is cross-reactivity among the allergens of hymenopteran venoms: it is greatest between yellowjacket and hornet venoms, as seen with Alec. There is limited cross-reactivity between venoms of paper wasps and other vespids (yellowjackets and hornets) and minimal cross-reactivity between honeybee venom and those of other hymenoptera.

Clinical reactions to insect stings can be divided into local, systemic allergic, systemic toxic, and delayed reactions. As noted in the introduction, local reactions and large local reactions to insect stings, arising mostly from nonspecific toxic effects of the venoms and in some cases from IgE-mediated reactions limited to the skin, are the most common. Systemic allergic reactions are most often IgE-mediated. Around 5–10% of patients experiencing large local reactions progress to a systemic reaction. In contrast, depending on concomitant risk factors, 25–70% of patients with a history of systemic reactions to stings will have systemic reactions to subsequent stings (Fig. 16.4). Risk factors for a systemic reaction to insect stings include age, preexisting cardiovascular or respiratory conditions, use of β-blocking agents and angiotensin-converting enzyme inhibitors, male gender, a history of previous anaphylaxis to insect stings, venom sensitization, an elevated serum tryptase above 11.4 ng ml^{-1}, and mastocytosis. As discussed in Case 20, mastocytosis is a proliferative disorder of mast-cell precursors resulting in excessive mast-cell accumulation, most commonly in the skin, bone marrow, liver, and spleen. The high systemic burden of mast cells results in elevated baseline levels of the characteristic mast-cell enzyme, tryptase.

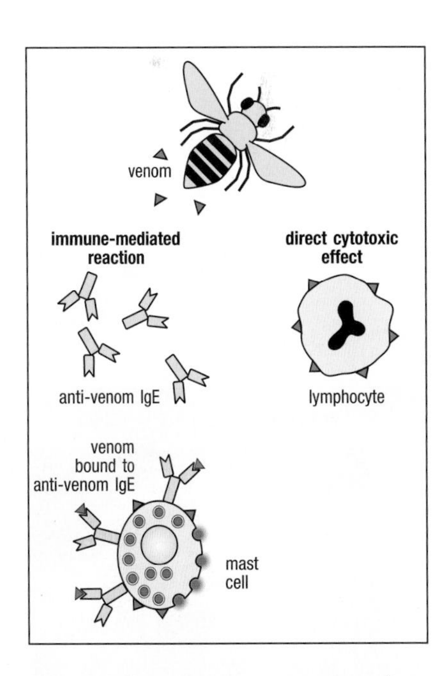

Fig. 16.3 Reactions to venoms can be IgE-mediated or non-IgE-mediated. IgE produced in response to the venom binds IgE receptors on mast cells. Binding of venom by these cell-associated antibodies triggers the mast cell to release histamine and other mediators. Venom can also interact directly with mast cells, triggering a similar response, or with lymphocytes, causing cytotoxic effects.

Local and systemic non-IgE-mediated reactions to insect stings are common. Fire-ant venom contains cytotoxic piperidine alkaloids, which cause a sterile pustule at the sting site (Fig. 16.5). Rarely, systemic toxic reactions are seen with multiple stings. The direct toxic effects of the large load of venom cause multiorgan dysfunction, including hemolysis, cardiac complications, acute renal failure, and rhabdomyolysis (breakdown of muscle fibers). Occasionally, delayed-onset reactions can develop days or weeks after the sting as a result of antigen–antibody complex formation and deposition in tissues. These include serum sickness, neuritis, encephalitis, vasculitis, and myocarditis.

The accurate diagnosis of an allergy to hymenopteran venom depends on taking a thorough clinical history complemented by testing for venom-specific IgE antibodies. The history may provide important clues regarding the culprit insect, such as the patient's recollection of the insect's appearance, a stinger

Reaction type	Age	Risk of systemic reaction (1–9 years after initial reaction)	Risk of systemic reaction (10–20 years after initial reaction)	Recommended clinical management
No reaction	Adult	17%		None
Large local	All	10%	10%	Supportive
Cutaneous systemic	<16 years	10%	5%	Supportive; no testing for venom hypersensitivity
	>16 years	20%	10%	Diagnostic testing for venom hypersensitivity; consider immunotherapy
Anaphylaxis	<16 years	40%	30%	Diagnostic testing for venom hypersensitivity; venom immunotherapy
	>16 years	60%	40%	Diagnostic testing for venom hypersensitivity; venom immunotherapy

Fig. 16.4 Risk of systemic reactions to stinging insects and recommended clinical management.

left in the skin (characteristic of honeybees), or the presence of pustule (found after fire-ant stings). However, it can be difficult to identify the culprit insect definitively, and skin testing against a panel of insects found in the patient's geographic location should be considered.

The decision to perform testing depends in part on the patient's age. Diagnostic testing is recommended for all individuals who have had a systemic reaction, including children such as Alec. However, skin testing is not necessary for children younger than 16 years old who have had only a cutaneous reaction (widespread hives without any other associated symptoms). These children have only a 10% risk of a systemic reaction to subsequent stings; for those who do experience a subsequent systemic reaction, there is less than a 1% risk of anaphylaxis.

Both skin testing and *in vitro* immunoassays for venom-specific IgE can be used to diagnose venom hypersensitivity, but skin testing is the preferred initial approach. The sensitivity of skin-prick testing, in which the skin is scratched and a drop of allergen extract is applied at the surface, is lower than that of intradermal testing, in which allergen is injected, for the diagnosis of venom hypersensitivity. Thus, unlike other types of allergies, intradermal testing is recommended as part of the initial work-up. To avoid false-negative results associated with a 'refractory period' after insect-sting reactions, skin testing should be performed at least 3–6 weeks after the sting. Skin testing yields positive results in more than 65% of patients with a history of systemic reaction to hymenopteran stings.

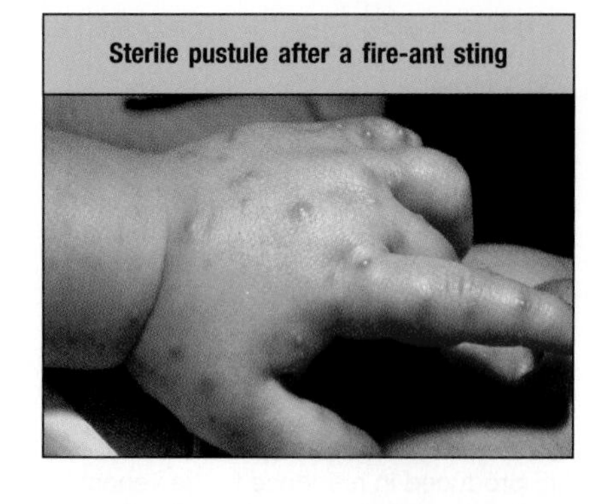

Fig. 16.5 Sterile pustule after a fire-ant sting. Unlike the venoms of bees and wasps, that of the imported fire ant has a very low protein content, and so tends to cause only a very local nonimmunologic reaction. Photograph courtesy of the Texas Department of Agriculture.

A negative skin test in patients with severe systemic reactions should be further evaluated with *in vitro* testing for venom-specific IgE, as approximately 10–20% of patients with a negative skin test will test positive for anti-venom IgE antibodies. In patients with no detectable venom-specific IgE antibodies, skin and *in vitro* testing should be repeated within 3–6 months. Negative test results from patients with systemic reactions cannot definitively exclude the possibility of systemic reactivity to subsequent stings, although such reactivity is rare. In patients with a history of severe systemic reaction but no detectable venom-specific IgE antibodies, baseline serum tryptase should be determined, to exclude mastocytosis.

Approximately 1–8% of patients presenting with venom hypersensitivity have an underlying clonal mast-cell disease, mastocytosis. These patients are at a high risk of developing severe systemic reactions. In patients without detectable IgE antibodies against venom, insect stings can cause mast-cell activation spontaneously or through non-IgE-mediated mechanisms. A tryptase test was done for Alec in the emergency room because the sting had occurred less than an hour earlier. Tryptase is released by activated mast cells and can be detected in the circulation up to about 4 hours after an anaphylactic event. Alec's allergist ordered a baseline tryptase level 1 month after the anaphylactic episode to test for mastocytosis. Because the level was less than 11.4 ng ml^{-1}, mastocytosis was excluded as a possible diagnosis in Alec's case.

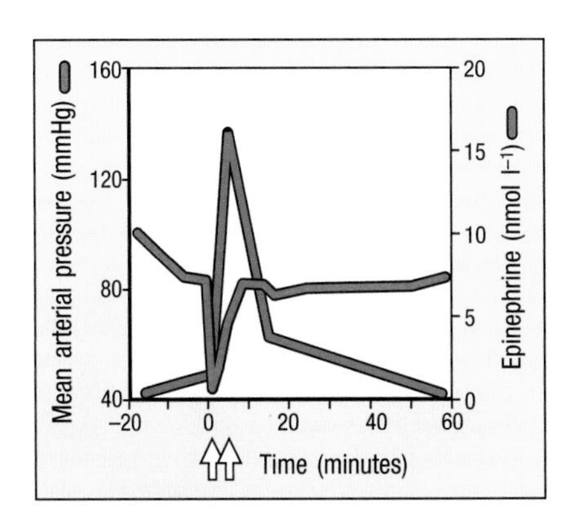

Fig. 16.6 Mean arterial pressure and epinephrine levels in a representative patient with insect-sting anaphylactic shock. Time 0 indicates the onset of the anaphylactic reaction as reported by the patient. The arrows indicate administration of antihistamines and epinephrine.

The treatment of local reactions to insect stings is supportive, with cold compresses, antihistamines, and sometimes oral steroids in severe cases. Intramuscular epinephrine is the immediate treatment for an anaphylactic reaction to an insect sting (Fig. 16.6). It should be noted that generalized urticaria without any other symptoms is not considered as anaphylaxis in children under the age of 16 years because cutaneous manifestations do not predictably progress to anaphylaxis in this age group. All patients with venom hypersensitivity should carry an epinephrine autoinjector. After a systemic sting reaction, patients should be referred to an allergy specialist for diagnostic evaluation and possible venom immunotherapy.

Allergen immunotherapy is effective in reducing the risk and severity of subsequent reactions. As described in Case 3, allergen immunotherapy has multiple mechanisms, including the generation of $CD4^+$ $CD25^+$ regulatory T lymphocytes, a shift from T_H2 to T_H1 responses, increases in inhibitory cytokines such as IL-10, and increases in allergen-specific blocking antibodies of the IgG isotype. However, a decrease in allergen-specific IgE levels does not correlate with the efficacy of immunotherapy.

The decision to start venom immunotherapy (VIT) depends on both the clinical history and the age of the patient (see Fig. 16.4). Patients with a history of large local reactions are at low risk of a systemic reaction to future stings, and so are typically not considered candidates for VIT. Immunotherapy should be considered in all individuals with a history of systemic reaction to hymenopteran stings, namely reactions that involve more than one organ system (cutaneous, gastrointestinal, respiratory, cardiac, neurological). The most concerning reactions are those involving the respiratory or cardiovascular symptoms and it is very important that these patients receive VIT. Immunotherapy is not routinely recommended for children younger than 16 years old with generalized cutaneous reactions to insect stings, because they have a low risk of a systemic reaction if stung again. In contrast, VIT is typically offered to adults with generalized hives as the only manifestation of hypersensitivity.

Although there are many variations of immunotherapy, all regimens comprise 'build-up' and 'maintenance' phases. Build-up phases begin with a dose of 0.1–1.0 μg; on maintenance doses, both children and adults should ultimately receive 100 μg of each insect venom. For patients with positive diagnostic testing to multiple venoms, there is controversy over whether VIT should be instituted for all venoms against which IgE antibodies are detected, or only for a single venom if the insect that caused the sting can be identified. Conventional build-up schedules involve one injection per week and reach the maintenance dose in 3–6 months, depending on the starting immunotherapy dose and the occurrence of adverse reactions. During the maintenance phase, immunotherapy can be given every 4 weeks in the first year of VIT and then every 6–8 weeks afterwards. 'Accelerated build-up' schedules have the advantage of achieving the target therapeutic dose more rapidly than conventional schedules, with no higher incidence of systemic reactions observed with rush immunotherapy for inhalant allergens. Patients at very high risk for recurrent stings (for example beekeepers) and who suffered a severe initial reaction will benefit from accelerated build-up schedules.

VIT is generally well tolerated; however, it can be complicated by systemic reactions in up to 20% of individuals. Risk factors for systemic reactions to immunotherapy include the following: immunotherapy to honeybee venom; elevated baseline tryptase levels; concomitant use of β-blockers or angiotensin-converting-enzyme (ACE) inhibitors or a history of allergic rhinitis. In addition, patients are more likely to experience systemic reactions during the build-up rather than the maintenance stages of VIT. After a systemic reaction, it is common practice to move back to the dose that was previously tolerated. The decision on whether to continue immunotherapy after the occurrence of a severe systemic reaction must be made on an individual basis. Although cutaneous reactions can be reduced with antihistamines that act on the H1 receptor, these medications do not reduce the incidence of severe systemic allergic reactions.

In general, VIT should be maintained for at least 3–5 years, at which point 80–90% of patients will be protected from systemic reactions. There are no definitive criteria for discontinuing VIT, although studies have examined the predictive value of negative tests for anti-venom IgE antibodies and negative skin tests. No severe reactions have been reported in patients who have discontinued VIT after having negative skin and *in vitro* testing. However, negative skin tests occur in fewer than 25% of patients, even after 6 years of immunotherapy. Higher rates of relapse after VIT are associated with older age, a history of severe reactions to insect stings, honeybee allergy, systemic reactions during venom immunotherapy, elevated basal serum tryptase, mastocytosis, and a treatment duration of less than 5 years.

Questions.

1 *Propose a biological mechanism for negative skin tests for venom allergy and negative tests for venom-specific serum IgE in a patient with a convincing history of anaphylaxis to hymenopteran venom.*

2 *Design an experiment to show that the production of IL-10 contributes to immunological tolerance during VIT for bee venom.*

3 *Development of allergen-specific blocking antibodies is proposed as another mechanism of VIT. How do these antibodies result in immunologic tolerance?*

4 *Is there any correlation between a patient's clinical symptoms and the size of the reactions seen on skin testing or the level of IgE antibodies to venom?*

5 *Why do some patients have positive skin tests for multiple venoms? Are these patients truly allergic to all the venoms that give positive results?*

6 *Why are patients with mastocytosis at a higher risk for anaphylaxis in response to insect stings, even when they do not have detectable levels of venom-specific IgE antibodies?*

CASE 17 Hypereosinophilic Syndrome

A consequence of an overproduction of eosinophils.

Eosinophils are white blood cells characterized by a bilobed nucleus and the presence of abundant granules containing cationic proteins that interact with the negatively charged stain eosin (Fig. 17.1). They have a central role both in immune responses to infestation with multicellular parasites and in allergic inflammation (see Cases 3, 5, and 13). Eosinophils have two distinctive types of granules: specific and primary. The specific granules have an electron-dense matrix containing major basic protein (MBP) that is surrounded by a matrix of eosinophil peroxidase (EPO), eosinophilic cationic protein (ECP), and eosinophil-derived neurotoxin (EDN) (Fig. 17.2), all of which can damage parasites (Fig. 17.3), and normal tissue. Primary granules contain lysophospholipase, a protein that can crystallize, giving rise to Charcot–Leyden crystals, which are often found in the airways of asthmatic patients and at other sites of allergic inflammation. After eosinophil activation, lipid bodies are seen within the cells. These structures generate pro-inflammatory lipid mediators that include leukotrienes, prostaglandins, thromboxanes, and lipoxins.

Eosinophils arise in the bone marrow from a common hematopoietic progenitor committed to the production of basophils and eosinophils. Their differentiation is stimulated by the cytokines granulocyte–macrophage colony-stimulating factor (GM-CSF), interleukin-3 (IL-3), and IL-5. Mature

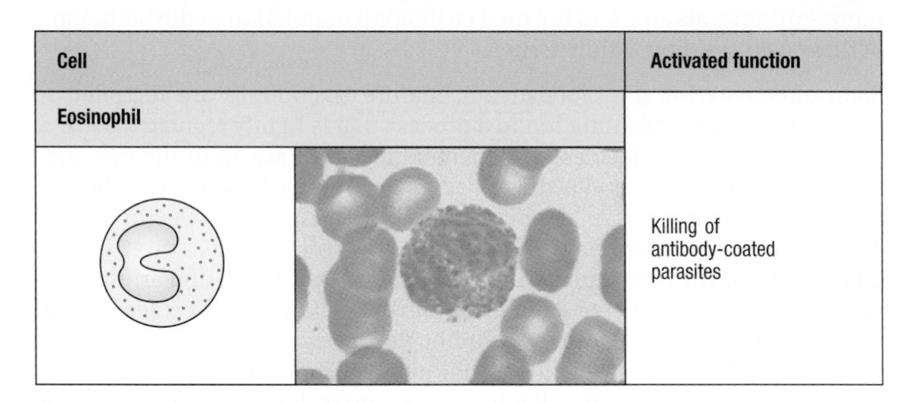

Cell
Activated function
Eosinophil
Killing of antibody-coated parasites

Fig. 17.1 The eosinophil. Eosinophils are granulocytes possessing eosin-staining granules that contain a variety of toxic and pro-inflammatory mediators; these are released on eosinophil activation. Eosinophils are thought to be involved in host defense against certain pathogens including fungi and parasites. However, they are also the cause of tissue damage in a variety of immune-mediated chronic allergic conditions, such as asthma and eczema. Photograph courtesy of N. Rooney, R. Steinman, and D. Friend.

This case was prepared by Hans Oettgen, MD, PhD, and Raif Geha, MD, in collaboration with Andrew MacGinnitie, MD, PhD.

Topics bearing on this case:

Migration and homing of leukocytes

Eosinophils and their functions

Fig. 17.2 Eosinophil products include toxic granule proteins and inflammatory mediators.

Class of product	Examples	Biological effects
Enzyme	Eosinophil peroxidase	Toxic to targets by catalyzing halogenation Triggers histamine release from mast cells
	Eosinophil collagenase	Remodels connective tissue matrix
	Matrix metalloproteinase-9	Matrix protein degradation
Toxic protein	Major basic protein	Toxic to parasites and mammalian cells Triggers histamine release from mast cells
	Eosinophil cationic protein	Toxic to parasites Neurotoxin
	Eosinophil-derived neurotoxin	Neurotoxin
Cytokine	IL-3, IL-5, GM-CSF	Amplify eosinophil production by bone marrow Eosinophil activation
	TGF-α, TGF-β	Epithelial proliferation, myofibroblast formation
Chemokine	CXCL8 (IL-8)	Promotes influx of leukocytes
Lipid mediator	Leukotrienes C_4, D_4, E_4	Smooth muscle contraction Increase vascular permeability Increase mucus secretion Bronchoconstriction
	Platelet-activating factor	Attracts leukocytes Amplifies production of lipid mediators Activates neutrophils, eosinophils, and platelets

eosinophils are nondividing cells, and maintaining a stable pool of eosinophils depends on regulated ongoing production in the bone marrow. Eosinophils released into the blood have a relatively short half-life (8–18 hours) and migrate rather quickly into tissues. The blood pool of eosinophils probably represents only about 1% of the total eosinophil population in the body, most of these cells residing within tissues.

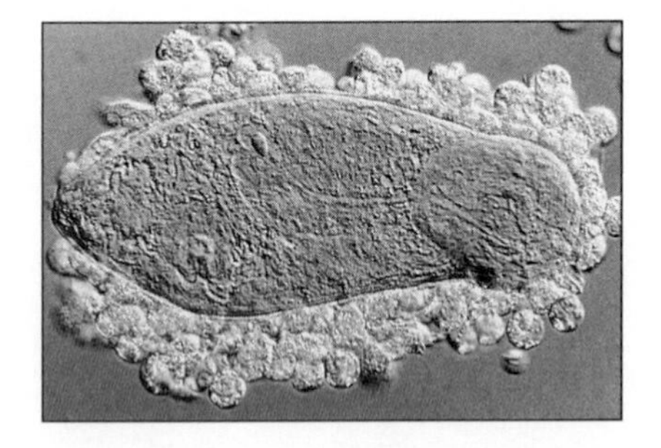

Fig. 17.3 Eosinophils attacking a schistosome larva. Large parasites, such as worms, cannot be ingested by phagocytes; however, when the worm is coated with antibody, eosinophils can attack it by binding via their Fc receptors for IgG and IgA. Such binding activates the eosinophils, which then release the contents of their toxic granules onto the surface of the parasite. Here, eosinophils are seen attacking a schistosome larva in the presence of antibody-containing serum from an infected patient. Photograph courtesy of A. Butterworth.

Soon after entering the bloodstream, mature eosinophils are recruited to peripheral sites of inflammation in a process that is tightly regulated by IL-5, vascular adhesion molecules, and chemokines (see Case 4). In the first step, eosinophils exit from free-flowing blood and roll along the surface of the vascular endothelium. This rolling interaction is mediated by P-selectin and the β_1 integrin VLA-4, which is expressed on activated endothelium at sites of inflammation. The rolling phase progresses to tight adhesion and arrest of the eosinophil in a process that depends on several adhesion molecules, including the CD18 family members lymphocyte function associated antigen-1 (LFA-1) and Mac-1 (which bind to endothelial intercellular cell adhesion molecule-1 (ICAM-1)), α_4:β_7 integrin (interacting with mucosal addressin cell adhesion molecule-1 (MAdCAM-1) on activated epithelium in the gastrointestinal tract), and VLA-4 (which binds to vascular cell adhesion molecule-1 (VCAM-1) and fibronectin). Both tight adhesion and subsequent emigration of eosinophils from vasculature to tissue are enhanced by the presence of small chemotactic proteins called chemokines, which signal via G-protein-coupled receptors on the eosinophil surface. CCR3 is the major chemokine receptor of eosinophils and interacts with CCL11 (eotaxin-1), CCL24 (eotaxin-2), and CCL26 (eotaxin-3).

Once they have moved into inflamed or infected tissue, eosinophils act by releasing the contents of their granules. The toxic proteins MBP, EPO, and ECP, as well as reactive oxygen species generated by the eosinophils, are harmful to parasites and tissue. Recent findings suggest that eosinophils, like neutrophils, activate a DNA trap mechanism to combat pathogens. Traps form when activated eosinophils eject mitochondrial DNA into the intracellular space. This DNA then combines with eosinophil granule proteins to ensnare pathogens. Cytokines and chemokines, including IL-5, released by tissue eosinophils promote the recruitment of additional eosinophils and other cells. IL-5 also sustains the survival of tissue eosinophils by inhibiting apoptosis (programmed cell death).

Tissue eosinophils can also contribute to the induction and regulation of immune responses. They express class II MHC molecules, which are responsible for antigen presentation to CD4 T cells. Activation of eosinophils leads to increased expression of MHC class II molecules on the cell surface and acquisition of antigen-presenting functions, leading to the activation of both naive and memory CD4 cells. Eosinophils promote the differentiation of CD4 T cells to T_H2 effector cells, leading to the production of IL-4 and IL-13. In addition, activated CD4 T cells produce IL-5, leading to a feedback loop that reinforces the generation of eosinophils.

Eosinophils are associated with allergic inflammation and are present in the affected tissues of patients with atopic dermatitis, allergic rhinitis, and allergic asthma. It is believed that eosinophils contribute to the persistence of inflammation in these illnesses. In addition, eosinophils are believed to have a role in the remodeling of the lung during chronic asthma. Remodeling is characterized by fibrin and collagen deposition, thickening of the basement membrane and increased smooth muscle mass in the affected lung. They can also be involved in nonallergic tissue damage. In the case discussed here, overproduction of eosinophils as a result of a somatic genetic defect leads to heart damage.

The case of Algernon Moncrieff: a middle-aged man with symptoms of heart damage caused by eosinophilia.

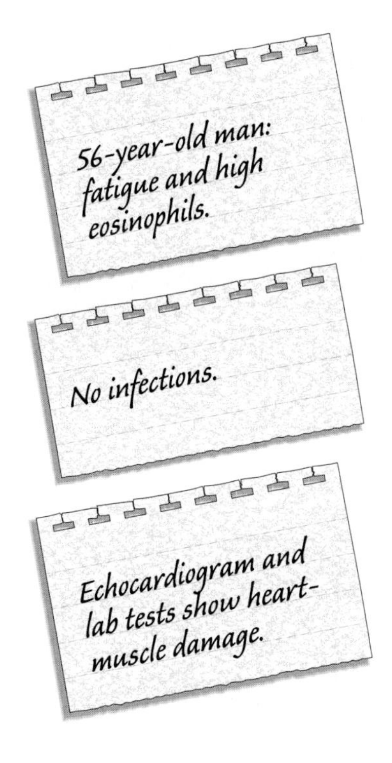

56-year-old man: fatigue and high eosinophils.
No infections.
Echocardiogram and lab tests show heart-muscle damage.

Algernon Moncrieff was a 56-year-old businessman who was referred to the allergy clinic after 2 years of increasing fatigue and decreasing tolerance of exercise. After several visits to his primary care doctor for these complaints, a blood sample was sent for analysis, including a complete blood count (CBC). The CBC was remarkable for an elevated eosinophil count of 6000 μl^{-1} (normal 0–500 μl^{-1}), representing 90% of the total white blood cell count (6700 μl^{-1}).

Mr Moncrieff told the allergist that he was on no medications or nutritional supplements and had not recently traveled. He then underwent several tests to investigate the possible causes of the eosinophilia. These included an HIV test, examination of his stool for parasites and their ova, and serological testing for infection with the parasitic roundworm *Strongyloides stercoralis*. Bone marrow aspirate and biopsy were only remarkable for increased eosinophils.

Additional investigations included a chest X-ray (normal), electrocardiography, which showed abnormally decreased voltages, and echocardiography, which showed mild thickening of the ventricular walls. Levels of the cardiac muscle enzymes troponin and creatinine kinase (myocardial fraction) were mildly elevated. A biopsy of the myocardium showed infiltration of eosinophils but no fibrosis.

Genetics results:
PDGFRα-FIP1L1
fusion.

Specialized cytogenetic testing of cells from the blood and bone marrow revealed a deletion in chromosome 4q, yielding a fusion between the FIP1-like-1 (*FIP1L1*) and the platelet-derived growth factor receptor α (*PDGFRα*) genes, which produced an abnormal fusion protein with tyrosine kinase activity derived from PDGFRα.

Because of concern that Mr Moncrieff was at risk of permanent cardiac injury as a result of the damaging effects of the toxic contents of eosinophil granules, he was treated with corticosteroids, which led to a rapid drop in eosinophil count to 200 within 48 hours of starting treatment.

The steroids were slowly tapered off, and long-term therapy was started with the tyrosine-kinase inhibitor imatinib (Gleevec). This led to a sustained decrease in Mr Moncrieff's eosinophil count into the normal range and a marked improvement in his symptoms. Eighteen months later his eosinophil count remained normal and he was asymptomatic with daily oral imatinib therapy.

Hypereosinophilic syndrome.

The causes of hypereosinophilia (elevated numbers of blood eosinophils) fall into three categories: reactive, hematopoietic malignancy, and hypereosinophilic syndrome (HES)—the last one being the cause of Mr Moncrieff's eosinophilia.

Reactive eosinophilia can be caused by several extrinsic causes, including infection with multicellular parasites such as the roundworms *Strongyloides* and *Toxocara* (Fig. 17.4) and reactions to medications, both of which are much more common than HES. For this reason a detailed history of both travel and the use of medicines, supplements, and other remedies is critical in evaluating patients with eosinophilia. Allergic diseases such as asthma and allergic rhinitis can be associated with eosinophilia, but they usually do not induce an eosinophil count above 1500. Adrenal failure and infection with HIV can also induce eosinophilia.

Hypereosinophilia can accompany a variety of hematologic cancers, including lymphomas and leukemias. The eosinophils may be either part of the malignant clone, as in eosinophilic leukemia, or reactive as in Hodgkin's lymphoma. Rarely, eosinophils are increased as a reaction to a nonhematologic malignancy such as adenocarcinoma.

HES is a diagnosis of exclusion, meaning that underlying causes of reactive eosinophilia and malignancy must be evaluated and ruled out before HES can be diagnosed. It can be divided into two subtypes, lymphocytic and myeloproliferative, on the basis of the identity of the causative cell population (Fig. 17.5).

Lymphocytic HES is characterized by a proliferation of aberrant T cells, which produce large amounts of hematopoietic cytokines. These T cells often exhibit an abnormal pattern of expression of cell-surface markers as well as markers of activation. The IL-5 produced by these cells leads to increased eosinophil production and survival. Note that the eosinophilia in lymphocytic HES is secondary to the T-cell abnormality. In contrast, myeloproliferative HES is due to an abnormal intrinsic proliferation of eosinophils and sometimes of other cells. This was the type seen in Mr Moncrieff. In most cases of myeloproliferative HES, as in his, there is a deletion in chromosome 4q12, leading to the production of a fusion protein between PDGFRα and FIP1L1. This fusion protein is a constitutively active tyrosine kinase. Many tyrosine kinases transmit growth signals to cells, and this overactive fusion protein is the proximate cause of the eosinophilia in myeloproliferative HES. Because myeloproliferative HES, at

Parasites associated with eosinophilia and specific organ involvement		
Organ involvement and parasite	**Exposure**	**Geographic distribution**
Gastrointestinal		
Hookworm*,†	Soil	Worldwide
*Ascaris**	Unpurified water, raw fruits, and vegetables	Worldwide
*Trichuris**	Unpurified water, raw fruits, and vegetables	Tropical
Anisakis	Raw fish	Worldwide
Heterophyes	Raw fish	Middle East, Asia
Capillaria	Raw fish	Asia
Liver		
Clonorchis	Raw fish and seafood	Asia
Opisthorchis	Raw fish and seafood	Asia
Schistosoma japonicum‡	Freshwater swimming	Asia
Schistosoma mansoni‡	Freshwater swimming	Latin America, Africa, Middle East
Fasciola	Watercress	Worldwide
Toxocara canis/Toxocara cati	Dogs, cats, soil	Worldwide
Lung		
Paragonimus†,‡	Crabs and crayfish	Asia
Ascaris	Unpurified water, raw fruits, and vegetables	Worldwide
Strongyloides†,‡	Soil	Tropical
Brugia malayi	Insect bite	Asia
Wuchereria bancrofti	Insect bite	Tropical
Toxocara canis/Toxocara cati†,‡	Dogs, cats, soil	Worldwide
CNS		
Angiostrongylus	Raw seafood	Asia
Gnathostoma†	Raw fish and poultry	Asia
Bladder		
Schistosoma haematobium	Freshwater swimming	Africa, Middle East
Muscle		
Trichinella‡	Pork	Worldwide
Eye		
Loa loa†	Insect bite	Africa
Onchocerca	Insect bite	Africa
Toxocara canis/Toxocara cati	Dogs, cats, soil	Worldwide
Lyphedema		
Wuchereria bancrofti	Insect bite	Tropical

Fig. 17.4 Parasitic infections associated with eosinophilia. CNS, central nervous system. *Usually diagnosed by examination of stool for ova and parasites. †May be associated with transient rash. ‡Serologic test available at the US Centers for Disease Control and Prevention.

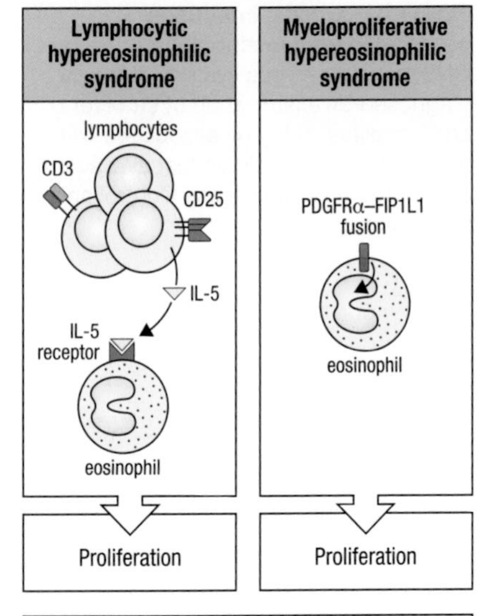

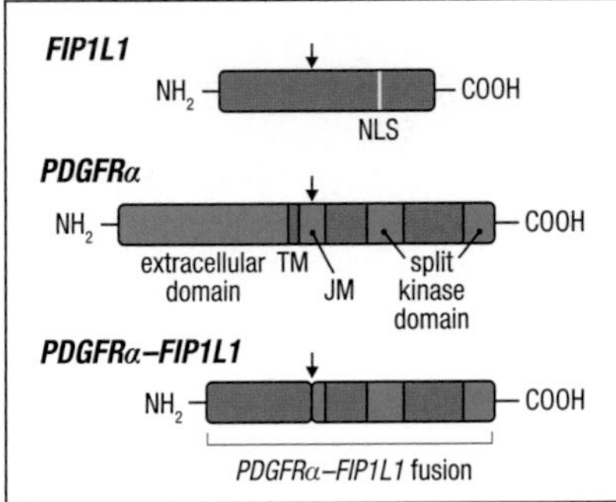

Fig. 17.5 Lymphocytic and myeloproliferative hypereosinophilic syndromes. In lymphocytic hypereosinophilic syndrome (upper left panel), an abnormal expansion of T cells (CD3-positive), typically bearing markers of activation (CD25) leads to the production of cytokines, including IL-5, that drive eosinophilopoiesis. In myeloproliferative hypereosinophilic syndrome (upper right panel), a somatic genetic alteration in a clone of progenitor cells provides a constitutive activating signal (arrow) leading to cell-autonomous proliferation. In the most common variety, the alteration is a fusion of the amino terminus of the *FIP1L1* gene with the carboxy terminus of the *PDGFRα* gene (lower panel). NLS, nuclear localization signal; TM, transmembrane region; JM, juxtamembrane region.

least in the case of PDGFRα-FIP1L1 rearrangement, is a clonal disorder, some immunologists believe it should be characterized as a chronic eosinophilic leukemia.

Eosinophils produce several toxic mediators that can damage normal tissue. Patients with HES can suffer injury to several organs, including the bone marrow, skin, heart, gastrointestinal tract, and nervous system. Patterns of organ involvement are variable. Cardiac involvement is common and begins with eosinophilic infiltration, which can be asymptomatic or cause inflammation and necrosis. The latter can lead to fibrosis and decreased cardiac function as well as intracardiac thrombus formation (Fig. 17.6).

Treatment to decrease eosinophil levels is generally recommended to minimize the chance of end-organ damage. However, some patients carry elevated eosinophil burdens for long periods without evidence of tissue damage. Before the advent of better-targeted treatments, corticosteroids were the mainstay of therapy for HES. Unfortunately, as with other immune-mediated diseases, prolonged corticosteroid use is associated with several adverse effects, including susceptibility to infection, osteopenia, and suppression of endogenous cortisol synthesis.

With recent advances in understanding the basis of HES, better-targeted therapies have become available. For the lymphocytic HES variant, blocking the effect of IL-5 with mepolizumab, an anti-IL-5 monoclonal antibody, is effective in decreasing eosinophil counts in many patients.

Similarly, inhibition of the PDGFRα-FIP1L1 fusion protein kinase activity using the tyrosine kinase inhibitor imatinib (Gleevec) is effective in patients with myeloproliferative HES resulting from this gene fusion. Interestingly, some patients who lack the *PDGFRα-FIP1L1* fusion gene also respond to imatinib. This suggests that their eosinophils may contain other genetic changes resulting in activated tyrosine kinases. A substantial number of patients with HES do not have a defined fusion gene and do not respond to imatinib. Other tyrosine kinase inhibitors are under development and may be useful for some of these individuals.

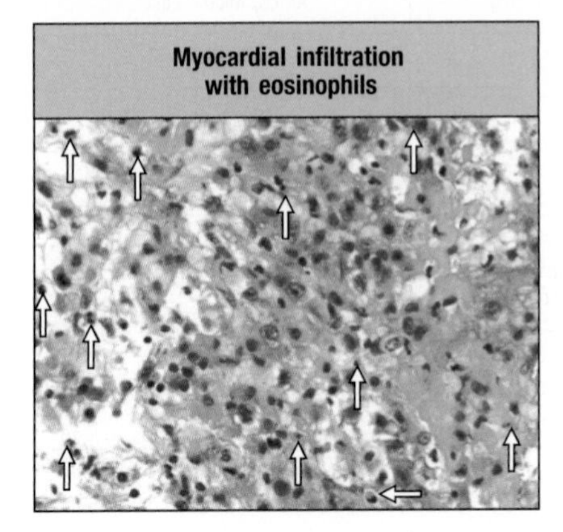

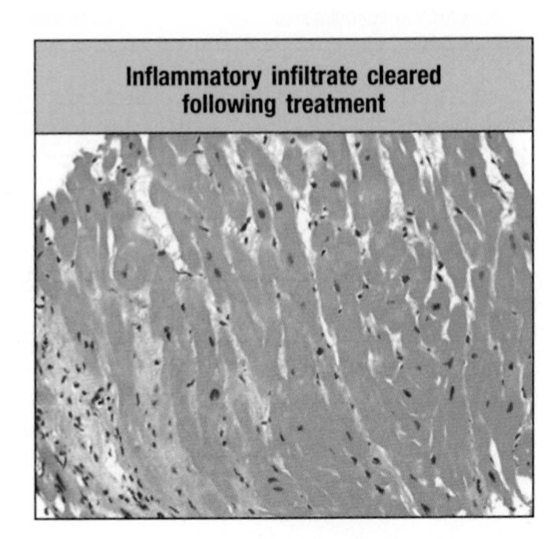

Fig. 17.6 Biopsies of myocardium in a patient with hypereosinophilic syndrome before and after treatment with corticosteroids. Left panel: microscopic examination of endomyocardial biopsy specimens (stained with hematoxylin and eosin) shows marked infiltration with eosinophils (arrows) and lymphocytes. Right panel: after treatment the inflammatory infiltrate has cleared, revealing normal cardiomyocytes.

Questions.

1 *Why is it important to treat eosinophilia promptly in patients with cardiac involvement?*

2 *Why might some people suffer early end-organ damage whereas others have asymptomatic eosinophilia for a prolonged period?*

3 *Most patients with HES are older adults. However, some pediatric cases of HES have been described. How might pediatric patients with hypereosinophilia differ from older patients?*

4 *Why is screening for Strongyloides stercoralis in particular recommended for patients with eosinophilia?*

5 *What other diseases are associated with activation of tyrosine kinases by genetic changes?*

6 *Mepolizumab is a monoclonal antibody directed against IL-5. What other diseases might respond well to mepolizumab?*

CASE 18 Churg–Strauss Syndrome

Granuloma formation, a rare complication of allergic asthma.

Granulomas are organized collections of macrophages, which often have an elongated (epithelioid) appearance. Granuloma formation serves to wall off intracellular pathogens and foreign material and is part of the normal immune response to these substances. In some instances, however, such as sarcoidosis, Crohn's disease, and vasculitis (destructive inflammation of blood vessels), granuloma formation can result from a misdirected and pathogenic pattern of inflammation.

The classic organism that induces granuloma formation is *Mycobacterium tuberculosis*. Mycobacteria can survive and grow intracellularly, even after phagocytosis by macrophages. The infected macrophages proliferate and aggregate to form the beginnings of a granuloma, recruiting additional macrophages and other immune-system cells. This aggregation serves to effectively wall off the infection and inhibit bacterial spread. Macrophage aggregation can lead to cell–cell fusion, creating large, syncytium-like structures appropriately named 'multinucleated giant cells.'

T lymphocytes have a critical role in orchestrating granuloma formation in response to *M. tuberculosis* infection (Fig. 18.1). Infiltrating T_H1-type $CD4^+$ T cells secrete tumor necrosis factor (TNF-α) and interferon (IFN)-γ, cytokines that drive the expansion and enlargement of the granulomas by activating macrophages (IFN-γ) and promote the recruitment of other effector cells to the site of infection (TNF-α). IFN-γ induces phagosome acidification along with a respiratory burst and production of reactive oxygen species, killing the intracellular mycobacteria. Although the granulomas found in *M. tuberculosis* infection are driven by a T_H1 response, granulomas in other infections may be T_H2-driven, or may exhibit a predominance of cell types besides $CD4^+$ T cells. Cytotoxic $CD8^+$ T cells, γ:δ T cells, and NKT cells can all contribute to cell killing in granulomas but are not necessary for granuloma formation. B cells can also be found within mycobacterial granulomas, but their contribution is less well delineated. Finally, fibroblasts are recruited to the site of inflammation and further help contain the bacterial spread by forming a protective shell.

Granulomas formed in response to mycobacterial infection often develop a necrotic center as the infected cells die. This type of granuloma is referred to as 'caseating,' as the necrotic material has the appearance of cheese (Latin

Topics bearing on this case:
Inflammation
Granuloma formation
Allergic asthma

This case was prepared by Hans Oettgen, MD, PhD, and Raif Geha, MD, in collaboration with Mindy Lo, MD, PhD.

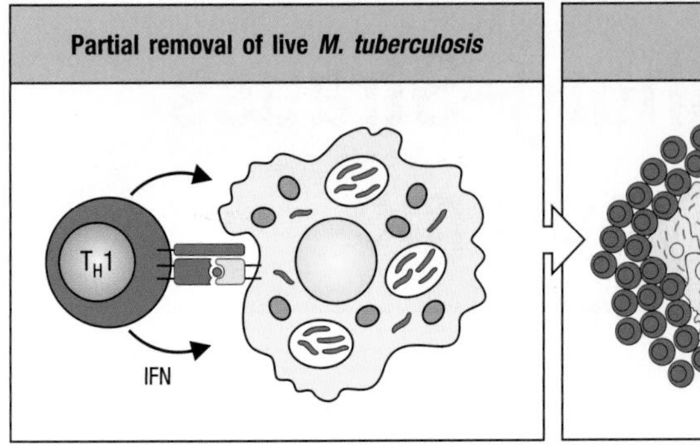

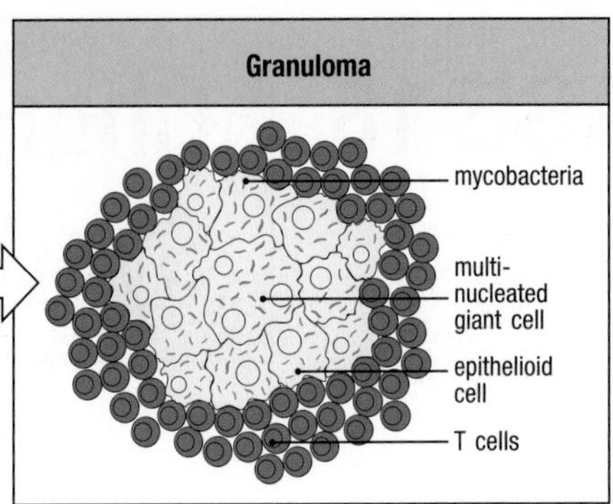

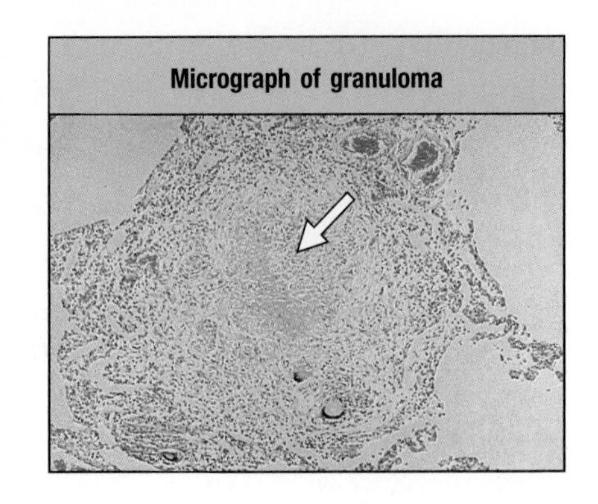

Fig. 18.1 *M. tuberculosis* infection and granuloma formation. Phagocytosis of intracellular mycobacteria (red) leads to antigen presentation to local T cells and then to T-cell-driven activation and aggregation of macrophages with a central core of infected cells. Multinucleated giant cells are formed by cell–cell fusion (left panel). Surrounding CD4⁺ T cells organize the macrophage response and induce the macrophage respiratory burst, killing intracellular organisms. CD⁺ T cells also secrete cytokines and chemokines to attract other effector cell types such as CD8⁺ T cells and NKT cells. Encapsulation of the infected cell mass by fibroblasts further limits mycobacterial dissemination. In some cases central necrosis (arrow) of the granuloma gives it the 'caseating' appearance (center panel). Granulomas can also occur in autoimmune diseases, including sarcoidosis (right panel), and in association with vasculitis, as in Churg–Strauss syndrome. Photograph courtesy of J. Orrell.

caseus). In contrast, granulomas related to autoimmune diseases, such as Crohn's disease, may not have caseating centers. Here, we look at a rare sequel to chronic allergic asthma that involves vasculitis and granuloma formation.

The case of Georgiana Darcy: neurologic symptoms in a patient with asthma.

42-year-old woman with chronic sinusitis and asthma.

At the age of 42 years, Georgiana Darcy attended the Children's Hospital rheumatology clinic because of a worsening of her symptoms of allergic rhinitis, with frequent sneezing and nasal congestion. Georgiana had begun to experience recurrent episodes of sinusitis in her early 30s; typically, the symptoms would be relieved by nasal saline rinses and antibiotics. A short time before her visit to the rheumatology clinic, Georgiana had been examined by an ear, nose, and throat surgeon who found that her nasal turbinates looked swollen and inflamed. He also noted a few small nasal polyps. He recommended a daily long-acting antihistamine, which did help her symptoms.

Georgiana had been diagnosed with asthma as a young woman, shortly after finishing college. These new symptoms had been attributed to the change in her environment, as she had moved to the East Coast after graduating. Her asthma was controlled with inhaled corticosteroids and a long-acting bronchodilator, and she would occasionally require brief courses of oral steroids for severe exacerbations.

Georgiana had recently noticed a tingling sensation in her right hand. She initially attributed this to carpal tunnel syndrome, as she spent most of the day working on a computer. However, the sensation grew worse over the next few weeks, and was accompanied by increasing fatigue and the appearance of a rash on her lower legs.

A physical examination revealed that Georgiana was breathing comfortably, although auscultation revealed faint wheezing. A neurologic examination showed decreased sensation over her right thumb and the index and middle fingers. She also had decreased strength in the right wrist. The doctor also noticed that Georgiana's gait was asymmetric, with a left-sided foot drop. The rash was in the form of a few dark-red lesions on her legs above the ankles.

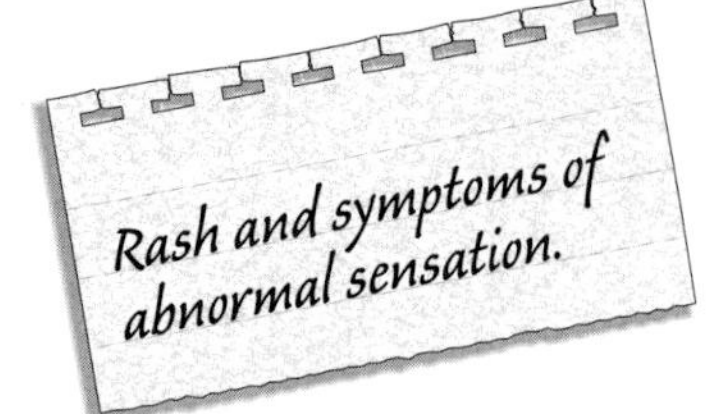

Laboratory tests revealed an elevated white blood cell count of 12,800 μl^{-1}, of which 50.1% were neutrophils, 10.4% lymphocytes, and 23.8% eosinophils. Georgiana's total eosinophil count was significantly elevated, at 3050 μl^{-1}. The erythrocyte sedimentation rate (ESR) was also high at 64 mm h^{-1} (normal less than 18), and C-reactive protein (CRP) was 6.9 mg dl^{-1} (normal less than 0.1), indicative of marked systemic inflammation. Georgiana tested positive for anti-neutrophil cytoplasmic antibody (ANCA), with confirmatory testing showing reactivity to the neutrophil cytoplasmic enzyme myeloperoxidase (MPO). Biopsy of one of the skin lesions showed a small-vessel

necrotizing vasculitis, with infiltrating eosinophils, neutrophils, and lymphocytes. In addition, the pathologist noted the presence of a granuloma, with several concentric layers of flattened (epithelioid) macrophages surrounding a core of eosinophilic debris in the dermis.

The diagnosis of Churg–Strauss syndrome was made, and Georgiana was admitted to the hospital and started on high-dose intravenous glucocorticoids. Within 3 days, there was a significant improvement in her rash, and a mild improvement in the tingling in her hand. Her ESR declined to 41 mm h^{-1}, and CRP also improved to 0.7 mg dl^{-1}.

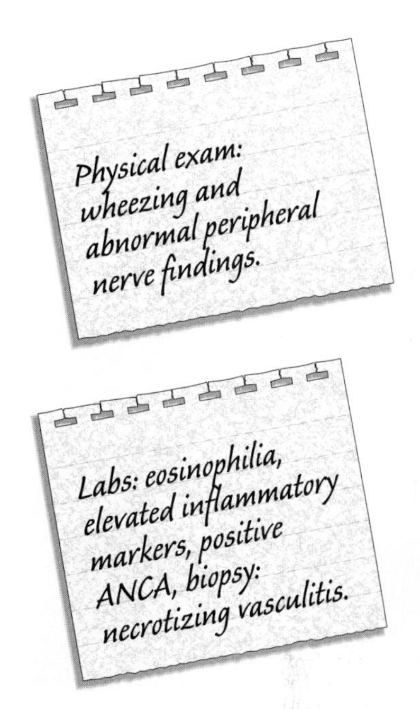

Physical exam: wheezing and abnormal peripheral nerve findings.
Labs: eosinophilia, elevated inflammatory markers, positive ANCA, biopsy: necrotizing vasculitis.

Churg–Strauss syndrome.

Churg–Strauss syndrome (CSS) is a rare form of vasculitis that primarily affects middle-aged patients with a previous history of adult-onset asthma. In 1951, two pathologists, Jacob Churg and Lotte Strauss, described a form of polyarteritis nodosa that was associated with allergic symptoms, increased peripheral eosinophils, and granulomatous lesions. They referred to this as allergic granulomatosis and angiitis, and it was subsequently renamed Churg–Strauss syndrome (Fig. 18.2).

Patients with CSS typically have a history of either adult-onset asthma or recently worsening asthma. This is considered the prodromal phase of the disease. There are often other signs of allergic disease as well, including rhinosinusitis and markedly elevated peripheral blood eosinophil count. Eosinophils are often increased in asthma, but counts are usually less than 1500 μl^{-1}; in CSS, in contrast, eosinophil counts can exceed 3000–5000 μl^{-1}. IgE is also usually significantly elevated. The pronounced eosinophilia is considered the second phase of the disease.

Progressive eosinophilia in CSS is followed by the development of both vasculitis and granulomas. Vasculitis in CSS can affect multiple organs, ranging from pulmonary manifestations to skin, neurologic, cardiac, and renal involvement. The neuropathy is caused by inflammation of the blood vessels that supply a nerve. Most commonly, a single nerve or a few large nerves are affected, a condition termed 'mononeuritis multiplex.' Skin signs are common and can range from purpuric lesions arising from vasculitis to larger subcutaneous nodules over the extensor surfaces of the arms and legs (Fig. 18.3), which represent the formation of granulomas. Cardiac involvement can be severe; biopsies of cardiac tissue reveal granulomatous nodules and interstitial eosinophilic

Criterion	Definition
Asthma	History of wheezing or diffuse high-pitched rales
Eosinophilia	> 10% eosinophils on white blood cell differential count
Mononeuropathy or polyneuropathy	Mononeuropathy, mononeuritis multiplex, or polyneuropathy attributable to vasculitis
Pulmonary infiltrates	Migratory or transitory pulmonary infiltrates on radiography attributable to vasculitis
Paranasal sinus abnormality	History of acute or chronic paranasal sinusitis or radiographic opacification of sinuses
Extravascular eosinophils	Biopsy showing blood vessel with accumulation of eosinophils in extravascular areas

Fig. 18.2 Diagnostic criteria for Churg–Strauss syndrome (CSS). The figure shows the American College of Rheumatology 1990 criteria for the classification of CSS. A diagnosis is suggested if four out of the six criteria are present.

Leukocytoclastic vasculitis rash

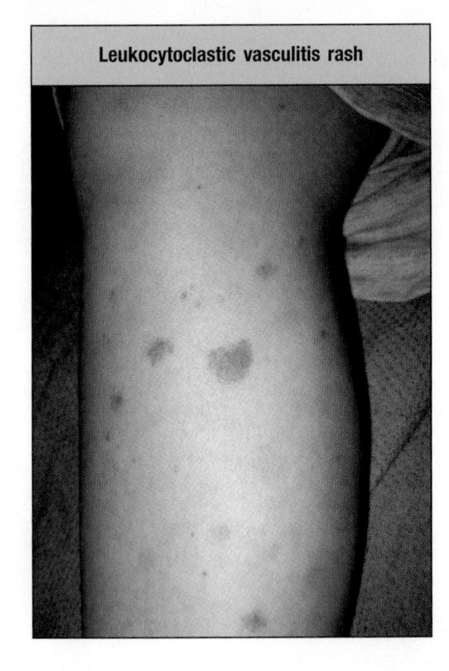

Fig. 18.3 Leukocytoclastic vasculitis rash. A patient with CSS showing the purpuric rash of leukocytoclastic vasculitis (LCV), a nonspecific term for a small-vessel vasculitis involving the skin. LCV can be associated with drug exposure, infection, or underlying autoimmune diseases. Histologically, lesions show perivascular neutrophil infiltration in the dermis. In CSS, this infiltration is dominated by eosinophils. Immune-complex deposition can be seen in or near the vessel walls; these deposits can contain the anti-neutrophil cytoplasmic antibody ANCA, which supports a role for this autoantibody in CSS pathogenesis. Photograph courtesy of M. Lee.

infiltration into the myocardium (Fig. 18.4), which can lead to arrhythmias, heart failure, and ultimately death.

The pathophysiology of CSS is incompletely understood. About half of patients with CSS test positive for ANCA. CSS is therefore considered a type of ANCA-associated vasculitis (AAV). Other AAV diseases include Wegener's granulomatosis (also known as granulomatosis with polyangiitis), microscopic polyangiitis, and polyarteritis nodosa. The ANCA antibody is classified as either cytoplasmic or perinuclear according to its immunofluorescence staining pattern in neutrophils. Cytoplasmic ANCA (c-ANCA) is typically directed against proteinase 3 (PR3), whereas perinuclear ANCA (p-ANCA) generally recognizes MPO. Both PR3 and MPO are components of neutrophil granules. CSS is more often associated with p-ANCA, whereas Wegener's granulomatosis is usually associated with c-ANCA.

There is some evidence that ANCAs are directly involved in pathogenesis. Although granular components are normally intracellular, stimulation of neutrophils by cytokines such as TNF-α causes increased localization of MPO and PR3 on the cell surface. Binding of ANCA to MPO or PR3 then triggers further neutrophil activation and degranulation. Degranulation releases superoxide radicals and proteases in a 'respiratory burst' that is toxic to neighboring cells. Because activated neutrophils also upregulate adhesion molecules that increase their attachment to vascular endothelial cells, this ANCA-triggered oxidative burst can lead to direct damage to vessel walls (Fig. 18.5).

There are several hypotheses about how CSS is triggered. Genetic factors seem to play a part, as association with HLA has been described (*HLA DRB4*-related alleles). Particular alleles of other genes, for example *CTLA4* and *IL10*, have also been associated with CSS. Environmental factors are also likely to contribute to the development of AAV. For example, one hypothesis is that exposure to silica dust, as can occur in construction workers exposed to blasting or jack-hammering, induces neutrophil activation and release of MPO. The MPO is then taken up by macrophages, which in turn present it as antigen to lymphocytes, triggering an autoimmune response. Infectious triggers may also be involved: *Staphylococcus aureus* has been reported to stimulate neutrophils directly. There are similarities between *S. aureus* peptides and those from PR3, suggesting that molecular mimicry might trigger autoimmunity in AAV. There are also speculations that some medications such as leukotriene inhibitors and omalizumab (a monoclonal antibody against IgE), both used to treat asthma, trigger CSS. However, it is likely that some patients have undiagnosed CSS that is unmasked when these agents are given in the setting of reducing glucocorticoid therapy.

Chest CT scan of a patient with CSS

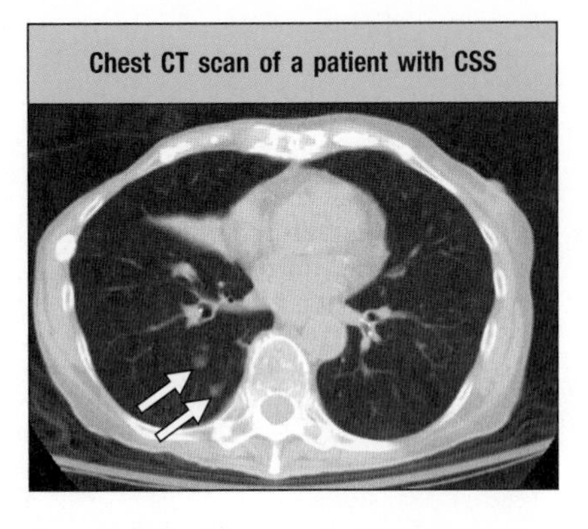

Fig. 18.4 Chest CT scan of a patient with CSS. Arrows indicate pulmonary nodules that represent granuloma formation. Photograph courtesy of G. Tsokos and V. Kyttaris.

Treatment of CSS and other AAVs is primarily immunosuppressive, with the use of systemic glucocorticoids and other immunosuppressant agents such as cyclophosphamide, azathioprine, and methotrexate. B-cell-depleting therapy using rituximab (a monoclonal antibody directed against CD20, a B-cell marker) has also been tried with some success in ANCA-associated vasculitis.

Questions.

1 *What is the significance of Georgiana's age at the time of her diagnosis of asthma?*

2 *What cytokine-directed therapy would you consider for Georgiana?*

3 *What might Georgiana's pulmonary function tests show?*

4 *How does Georgiana's presentation differ from that of hypereosinophilic syndrome?*

5 *Why do you think rituximab (anti-CD20 antibody) works as a treatment for Churg–Strauss disease?*

Eosinophilic vasculitis in a patient with CSS

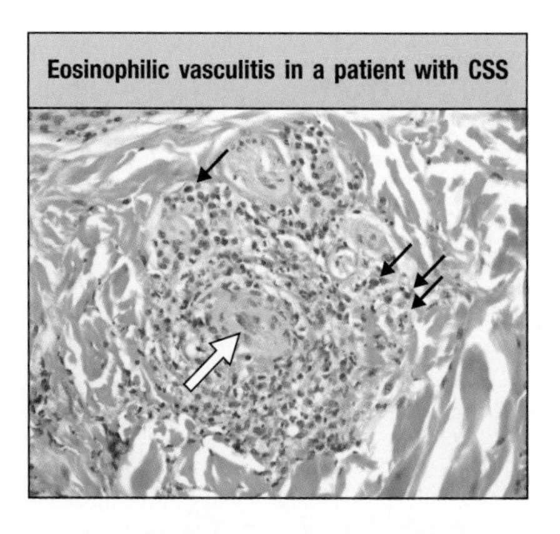

Fig. 18.5 Histopathology of a skin biopsy in a patient with Churg–Strauss syndrome. A chronic infiltrate rich in eosinophils (some indicated by small black arrows) surrounds a small blood vessel in this biopsy from a patient with Churg–Strauss Syndrome. The lumen of the vessel is obliterated by a pink-staining thrombus (large white arrow). Photograph courtesy of E. A. Morgan.

CASE 19 Aspirin-Exacerbated Respiratory Disease (AERD)

Lipid mediators of inflammation.

Arachidonic acid is a key inflammatory intermediate that is liberated from the phospholipids of cell membranes by the enzyme cytosolic phospholipase A_2 ($cPLA_2$) (Fig. 19.1). The lipid mediators formed from arachidonic acid include the prostanoids, which are produced via the cyclooxygenase pathway, and the leukotrienes, which are made by the lipoxygenase pathway. Prostanoids and leukotrienes are collectively called eicosanoids (derived from the Greek word for 20, *eicosa*) because of their 20-carbon chain. Eicosanoids can either propagate or inhibit inflammation, depending on the specific mediators and receptors engaged.

The enzyme 5-lipoxygenase converts arachidonic acid into the unstable leukotriene precursor LTA_4. LTA_4 is either converted into LTB_4 by the enzyme LTA_4 hydrolase, or into LTC_4 by LTC_4 synthase (see also Fig. 2.7). Outside the cell, LTC_4 is converted enzymatically into LTD_4 and LTE_4. LTC_4, LTD_4, and LTE_4 are collectively referred to as cysteinyl leukotrienes (Cys-LTs), because of the presence of the amino acid cysteine in their structure. The primary known function of LTB_4 is the recruitment of leukocytes to the sites of inflammation. The Cys-LTs are the most potent known constrictors of human airways and can also mediate increases in vascular permeability, mucus production, eosinophil recruitment, and mast-cell proliferation. Unlike LTC_4 and LTD_4, LTE_4 accumulates in both plasma and urine, and so the urinary LTE_4 level can be used as a marker of Cys-LT production. Two types of drugs that antagonize the function of leukotrienes are used in the treatment of asthma and allergic disease (see Fig. 2.7). Zileuton inhibits leukotriene production by blocking 5-lipoxygenase, whereas the drugs montelukast and zafirlukast block the main Cys-LT receptor $CysLT_1$ but not the other known Cys-LT receptor, $CysLT_2$.

Cyclooxygenase (COX) enzymes catalyze the formation of prostanoids from arachidonic acid. There are two main COX enzymes, COX1 and COX2, which are encoded by separate genes and have different patterns of expression. COX1 is constitutively expressed in most tissues, whereas COX2 is inducible and transiently active in tissues in the context of inflammation. The prostaglandin PGE_2 is the main prostanoid produced in the airway epithelium and smooth muscle cells. It has anti-inflammatory properties, which include the suppression of the synthesis of Cys-LTs via the inhibition of 5-lipoxygenase.

Topics bearing on this case:

Prostaglandins and leukotrienes

Chronic asthma

This case was prepared by Hans Oettgen, MD, PhD, and Raif Geha, MD, in collaboration with Ari Fried, MD.

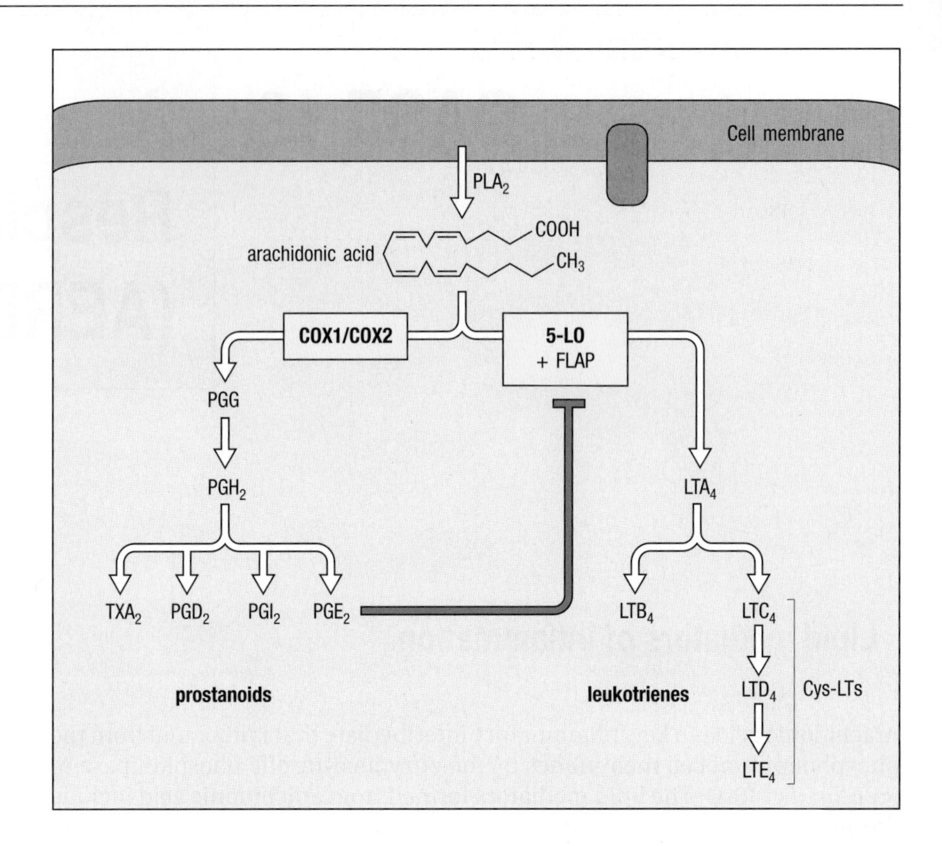

Fig. 19.1 Arachidonic acid metabolism. Lipid mediators are synthesized from arachidonic acid: the prostanoids through the cyclooxygenase (COX) pathway, and the leukotrienes through the 5-lipoxygenase (5-LO) pathway. Phospholipase A_2 (PLA_2) catalyzes the liberation of arachidonic acid from cell membranes. COX1 or COX2 metabolizes arachidonic acid to prostaglandin (PG) H_2, the common precursor for the main prostaglandins thromboxane A_2 (TXA_2), PGD_2, PGE_2, and PGI_2. 5-LO and the membrane protein 5-lipoxygenase-activating protein (FLAP) convert arachidonic acid to the leukotriene precursor LTA_4, which can be metabolized to LTB_4 or to the cysteinyl-leukotrienes LTC_4, LTD_4, and LTE_4. PGE_2 acts as a brake on leukotriene production through the inhibition of 5-LO and FLAP.

Aspirin (acetylsalicylic acid) and the nonsteroidal anti-inflammatory drugs (NSAIDs) exert their pain-relieving and anti-inflammatory actions through the inhibition of COX enzymes (Fig. 19.2). In this case we consider a hypersensitivity to aspirin that triggered the development of asthma in a young woman with no previous history of the disease.

The case of Verena Tarrant: treating a headache triggers a first severe asthma attack.

Verena: 18-year-old female with chronic nasal congestion.
Onset of asthma after aspirin ingestion. No history of asthma or atopy.

Verena Tarrant, an 18-year-old woman, was referred by her primary care physician to the Children's Hospital Allergy Clinic. She had been complaining of a runny and congested nose for the previous 2 years, and in recent months she had additionally developed a mild cough and expiratory wheezing without any apparent triggers. One evening she complained of a slight headache and symptoms of a common cold. Her mother gave her an over-the-counter cold remedy that contained aspirin. Instead of the anticipated relief, her cold symptoms became much worse with profuse watery drainage from her nose and eyes. Within an hour she had developed a dry cough, wheezing, and shortness of breath. By the time the family arrived at the local Emergency Room, Verena was in severe respiratory distress and unable to talk. Symptoms responded only minimally to three consecutive treatments with nebulized albuterol. She was transferred to the intensive care unit, where she received intravenous treatment with bronchodilators and corticosteroids. She stayed in the hospital for 5 days until her symptoms were controlled well enough to allow her to go home.

Verena was diagnosed with asthma, and treatment for asthma was prescribed with a daily dose of an inhaled corticosteroid. In the following year she had to stop playing sports because of frequent episodes of wheezing and coughing during practice. She

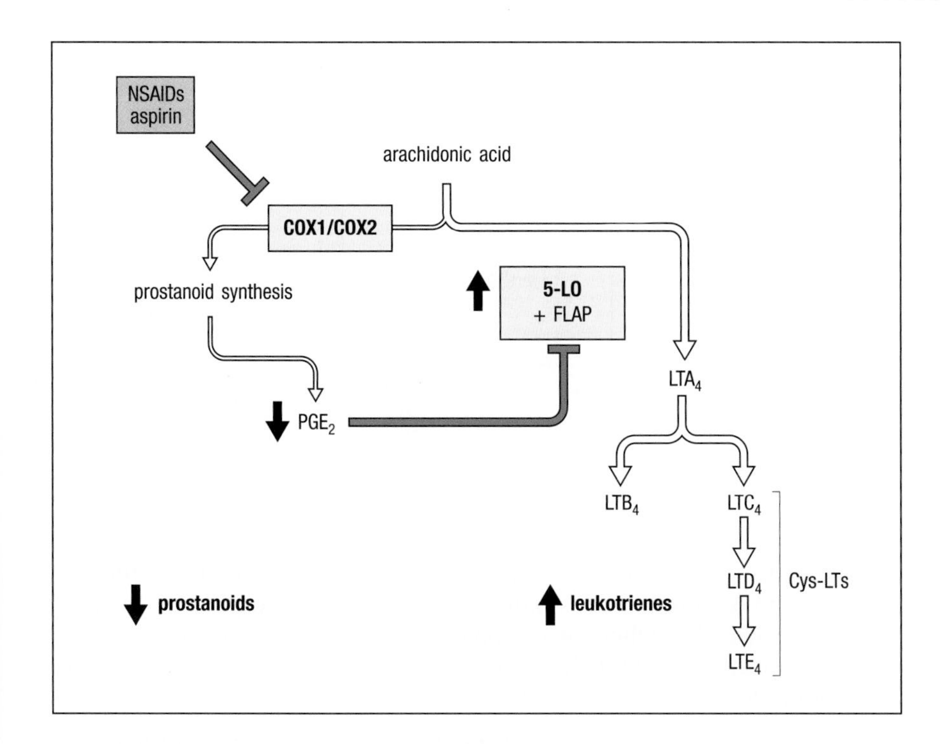

Fig. 19.2 Aspirin and cross-reacting NSAIDs inhibit COX enzymes. This inhibition decreases PGE_2 production and releases inhibition on the production of pro-inflammatory leukotrienes. Patients with AERD have low PGE_2 activity at baseline. Further inhibition by aspirin or NSAIDs results in unopposed leukotriene synthesis.

needed regular treatment with albuterol, and the dose of inhaled corticosteroid had to be increased.

Skin-prick testing did not identify any environmental allergies. Verena was found to have multiple nasal polyps, which occluded her nasal passages and sinuses. These were removed by endoscopic surgery. Relief of symptoms after surgery was short-lived because nasal polyps regrew within 6 months.

A supervised provocative challenge to aspirin was conducted to determine whether Verena was genuinely sensitive to aspirin. She was pretreated with the leukotriene antagonist montelukast to prevent severe lower respiratory symptoms during the challenge. Verena then received gradual increments of aspirin in the hospital and was closely monitored for symptoms. She developed profuse rhinorrhea, sneezing, and flushing of the skin 30 minutes after the second dose of aspirin and was treated symptomatically with antihistamines. This reaction confirmed her aspirin sensitivity. She was therefore advised to continue taking aspirin, following a protocol designed to desensitize patients. She tolerated subsequent aspirin doses without any further reactions and was instructed to continue taking aspirin twice daily at home. Her asthma, rhinitis, and nasal polyposis improved substantially over the following months.

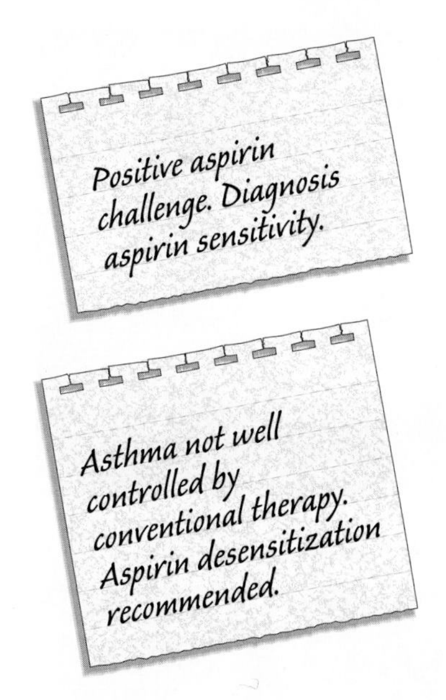

Positive aspirin challenge. Diagnosis aspirin sensitivity.
Asthma not well controlled by conventional therapy. Aspirin desensitization recommended.

Aspirin-exacerbated respiratory disease (AERD).

Hypersensitivity to aspirin and NSAIDs is a unique form of drug hypersensitivity caused by alterations in eicosanoid biosynthesis (see Fig. 19.2). Cultured cells from aspirin-sensitive individuals have low PGE_2 production at baseline. It has been hypothesized that low PGE_2 production *in vivo* might release Cys-LT synthesis from inhibition by the COX product PGE_2. When COX1 is inhibited by aspirin or NSAIDs, the synthesis of COX1 products, including PGE_2, decreases rapidly, which results in augmented synthesis of Cys-LTs. In susceptible individuals this leads to excessive Cys-LT synthesis, causing

Fig. 19.3 Classical clinical features of AERD.

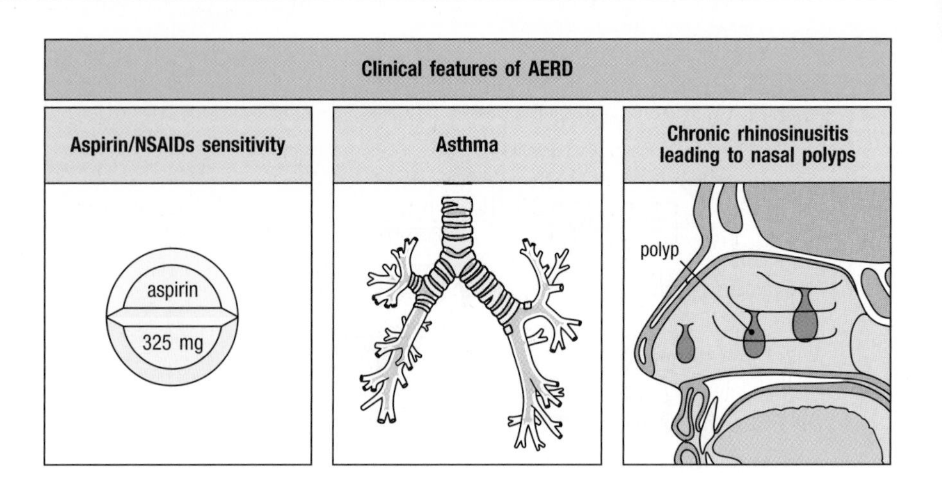

bronchospasm, release of inflammatory mediators such as histamine and tryptase, and recruitment of inflammatory cells to the respiratory mucosa. Aspirin-provoked bronchoconstriction in sensitive subjects can be prevented by the administration of inhaled PGE_2, directly demonstrating the role of this arachidonic acid product in aspirin sensitivity.

No anti-aspirin antibodies have been identified in patients with aspirin hypersensitivity, and skin-prick tests to aspirin are negative. Furthermore, patients can develop reactions after exposure to structurally dissimilar NSAIDs because of their common pharmacological effects on the COX pathway. Such broad cross-reactivity among different chemical agents is not seen in classic IgE-mediated allergies. Given its non-immunological underlying mechanism, this hypersensitivity, known as aspirin-exacerbated respiratory disease (AERD), is classified as a non-allergic hypersensitivity reaction.

Aspirin sensitivity, asthma, nasal polyposis, and chronic sinusitis are the hallmark clinical features of AERD (Fig. 19.3). Symptoms are rarely present in childhood and most often start between the teenage years and 40 years of age. Patients have intractable inflammation of the respiratory tract that usually begins as rhinitis and evolves into chronic rhinosinusitis with nasal polyps, followed by asthma. Aspirin sensitivity often becomes apparent only after other symptoms have manifested themselves. About half of patients with AERD have concomitant atopy with detectable sensitization to inhalant allergens, but this was not the case with Verena. The sinusitis and asthma associated with AERD often run a protracted course and respond poorly to standard therapies. Sinusitis is accompanied by the formation of nasal polyps, which grow large enough to result in anosmia, the loss of the sense of smell (Fig. 19.4). Polyps characteristically regrow after surgical removal, requiring repeat surgery every three years on average in patients with AERD. Asthma is severe in many patients, frequently necessitating treatment with systemic corticosteroids in addition to inhaled corticosteroids.

Nasal polyp

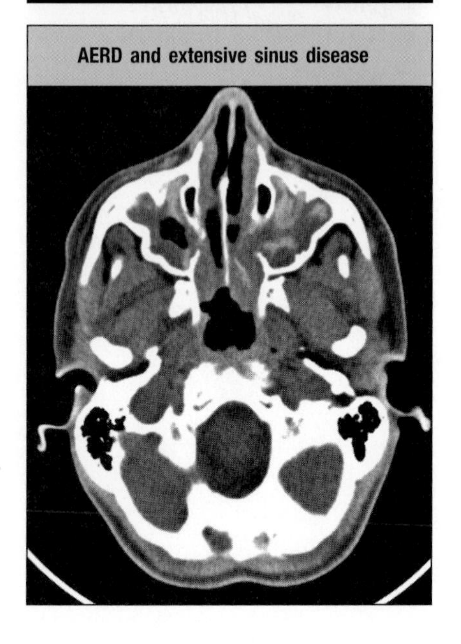

Fig. 19.4 The nasal cavities of a patient with AERD. The upper panel shows a nasal polyp. In this image a large yellow polyp is seen arising from the middle meatus of the nasal cavity. The polyp pushes the middle nasal turbinate medially. Nasal polyps are soft, noncancerous growths that develop in the side walls of the nasal passages as a result of chronic inflammation. In the lower panel a computed tomography scan shows a transverse section of the head of a patient with AERD and extensive sinus disease. The nose is seen at the top and is surrounded by the paranasal sinuses. There is streaky increased density and opacification of the maxillary sinuses, which represents chronic sinusitis with polyps. Polypoid opacities protrude from the paranasal sinuses into the nasal cavity.

Eosinophil numbers are markedly increased in the respiratory mucosa of patients with AERD as a result of the upregulation of inflammatory cytokines and lipid mediators. Most patients with AERD produce excessive amounts of Cys-LTs even without aspirin exposure, and these levels show further marked increases after exposure to nonselective COX inhibitors (those that act on both COX1 and COX2) such as aspirin. Exposure to nonselective COX inhibitors is not required to initiate disease, but once sensitivity has developed, exposure can precipitate violent reactions. These include bronchospasm, profuse rhinorrhea, conjunctival injection, flushing of the skin and variable skin eruptions, and periorbital swelling, and can sometimes resemble life-threatening anaphylaxis. It is believed that no single defect accounts for all cases of AERD but that pathologic underproduction or overproduction of different lipid mediators leads to the same clinical presentation.

Reactions are followed by a refractory period, in which patients can take aspirin or NSAIDs without developing further symptoms. This observation has given rise to the strategy of desensitizing aspirin-sensitive patients through continued daily or twice-daily administration of aspirin. Desensitization is performed by challenging patients to increasing doses of oral aspirin at intervals of 1.5 to 3 hours. Upper and/or lower respiratory tract reactions are expected to occur, but once a patient has reacted, further doses are usually tolerated without the development of additional symptoms. Desensitization is only temporary: patients become reactive to aspirin again after stopping daily treatment.

Interestingly, desensitization not only prevents the occurrence of reactions, but it also improves some of the clinical features and pathologic findings characteristic of AERD. Inflammation of the nasal airways diminishes significantly, nasal polyp growth diminishes, and asthma may improve, allowing a marked reduction in the use of corticosteroids to control disease. The mechanism underlying these changes is not well understood, but it possibly has to do with a correction of imbalances in arachidonic acid metabolism. Research studies have shown that NSAID-induced alterations in leukotriene synthesis and $CysLT_1$ receptor expression in the respiratory mucosa are normalized in patients after desensitization. Because of its known beneficial effects, desensitization is offered as a therapeutic option to patients with poorly controlled AERD.

Questions.

1 *Can you think of medications that could be used as alternatives to aspirin in patients with AERD?*

2 *Why do reactions to aspirin sometimes resemble anaphylactic reactions despite the absence of measurable IgE antibodies against aspirin?*

3 *What would you expect when you measure urinary LTE_4 levels in patients with AERD compared with healthy individuals at baseline and during an aspirin challenge?*

4 *Because overproduction of leukotrienes is thought to have a key role in aspirin-provoked reactions, could such reactions be prevented by taking leukotriene antagonists?*

CASE 20 Mastocytosis

A disorder of mast-cell proliferation and activation.

So far in this book, the activation of mast cells has been encountered mainly in the context of IgE-induced allergic reactions (see, for example, Case 1). However, mast cells can also be induced to release mediators in the absence of IgE. IgE-independent mast-cell activation is often evident in situations in which mast-cell numbers are markedly increased.

Mast cells develop in tissues from CD34$^+$ bone marrow-derived precursors, 'mast cell progenitors.' These cells, which do not have any recognizable features of mast cells, travel to tissues via the bloodstream and there differentiate into mature mast cells under the influence of local factors. The differentiation and proliferation of mast cells are induced by the interaction of the receptor tyrosine kinase Kit on the mast cell with its ligand, stem-cell factor (SCF), which is produced by cells residing in the mucosal and connective tissue sites. This interaction drives intracellular signals that result in mast-cell differentiation, growth, and protection from apoptosis. Certain mutations in the *KIT* gene cause the receptor protein to assume an active form even in the absence of its ligand (Fig. 20.1), leading to the excessive differentiation and proliferation of mast cells and their accumulation in the skin, bone marrow, and mucosal sites including the gastrointestinal tract—causing a disorder known as mastocytosis. In this disease, mast cells can be activated (Fig. 20.2) by a variety of non-immunological factors, or even in the absence of any discernible trigger. The consequent release of mediators contained in the mast-cell granules causes a wide variety of symptoms, as we shall see in this case.

Topics bearing on this case:

Mast-cell activation

This case was prepared by Hans Oettgen, MD, PhD, and Raif Geha, MD, in collaboration with Cem Akin, MD, PhD.

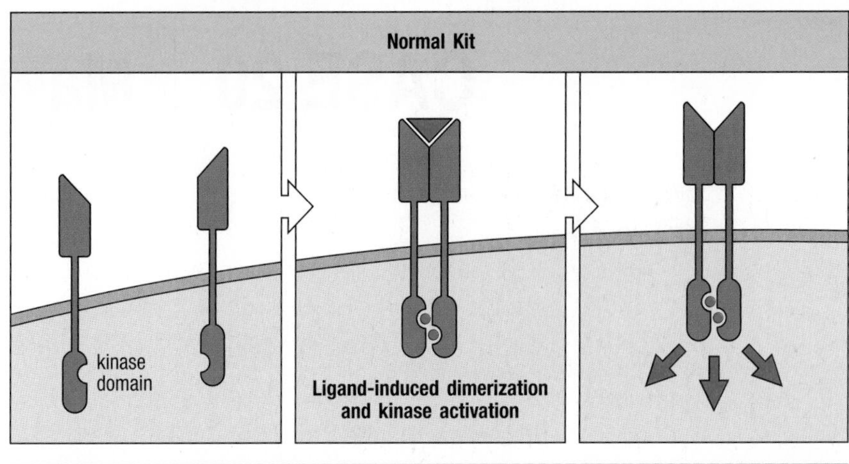

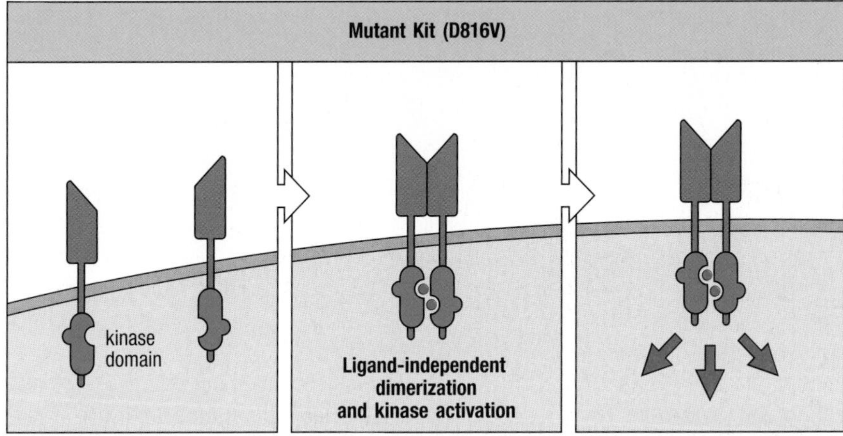

Fig. 20.1 The growth-factor receptor Kit controls mast-cell proliferation and differentiation. Top panel: Kit, the receptor for stem-cell factor (SCF), the major growth factor for mast cells, is a transmembrane protein with an extracellular ligand-binding domain and intracellular tyrosine kinase domains. Binding of SCF promotes receptor dimerization and kinase activation, leading to the phosphorylation of several tyrosine residues in the receptor and in downstream substrates. Lower panel: in mast cells harboring a mutation in the active site of the Kit tyrosine kinase domain, receptor dimerization and kinase activation occur in a ligand-independent manner, leading to constitutive mast-cell expansion (mastocytosis).

The case of Giles Winterbourne: a 35-year-old man with a fixed, hyperpigmented macular skin rash, flushing, and lightheadedness.

Giles came to the allergy clinic at the age of 35 years, complaining chiefly about a hyperpigmented macular skin rash. He had noted the onset of the rash on his upper thighs about 5 years previously, but initially thought it was freckles. However, the lesions became more numerous over the next few years, eventually involving much

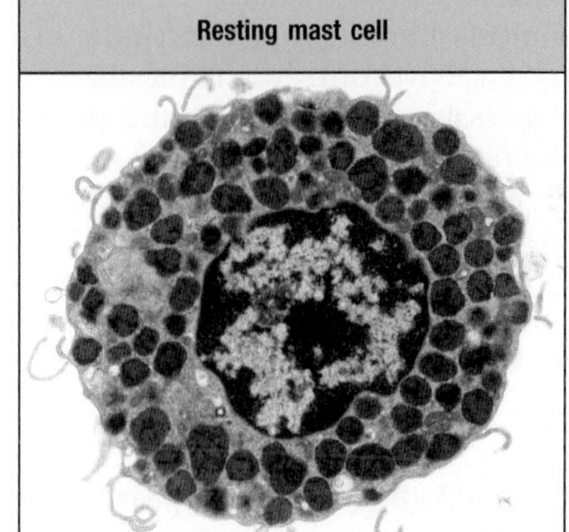

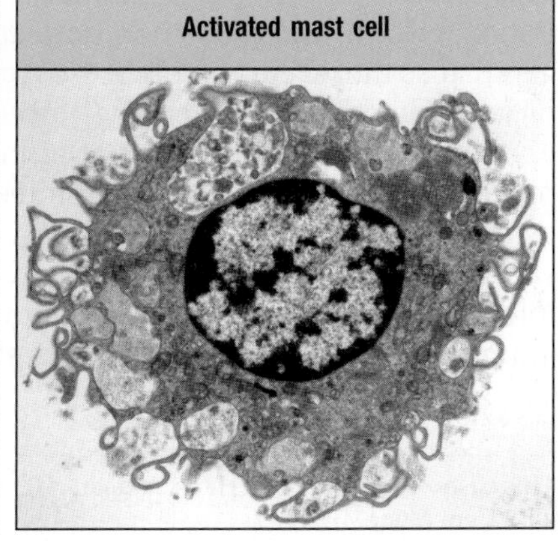

Fig. 20.2 Electron micrographs of resting and activated mast cells. Resting mast cells (left panel) are distinguished by their numerous large secretory granules—membrane-enclosed structures that contain many inflammatory and vasoactive mediators, including histamine. On activation, mast cells undergo degranulation (right panel); the granules are released by exocytosis, which liberates their contents to the exterior of the cell. Photographs courtesy of A.M. Dvorak.

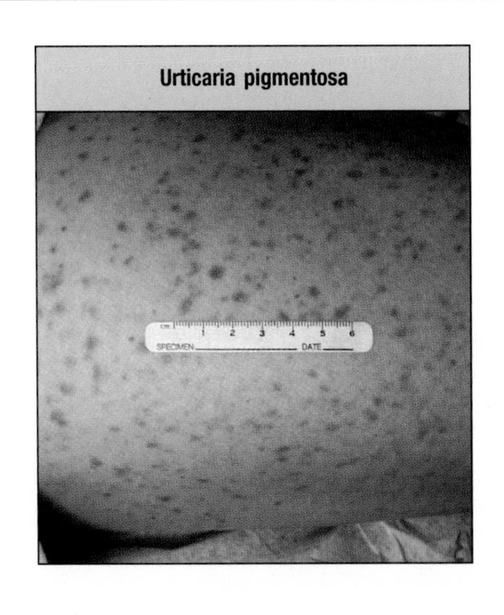

Fig. 20.3 Urticaria pigmentosa. The hyperpigmented maculopapular rash known as urticaria pigmentosa is shown here on the leg of a patient with mastocytosis. Photograph courtesy of K. Brockow.

of Giles's torso and limbs. The lesions were each a few millimeters in diameter and dark brown in color. They were not usually itchy but could flare up after a hot shower, if rubbed, or after exercise.

Giles had experienced episodes of facial flushing lasting for 10–30 minutes that were associated with tachycardia, lightheadedness, nausea, and abdominal cramping. He had almost fainted during some of these episodes. His cardiology evaluation did not show any evidence of cardiovascular disease, and testing for food allergies proved negative. Giles reported that he had about one or two of these episodes each month without any discernible triggers. He was asymptomatic between the episodes.

Giles's medical history was otherwise unremarkable. He did not have allergic rhinitis or asthma, and had no known allergies to medications, food, or latex. However, he had experienced itching on taking codeine after minor surgery and he remembered fainting after being stung by a honeybee 3 years earlier. On that occasion he was evaluated in the Emergency Room and was referred to an allergist, but did not keep the appointment.

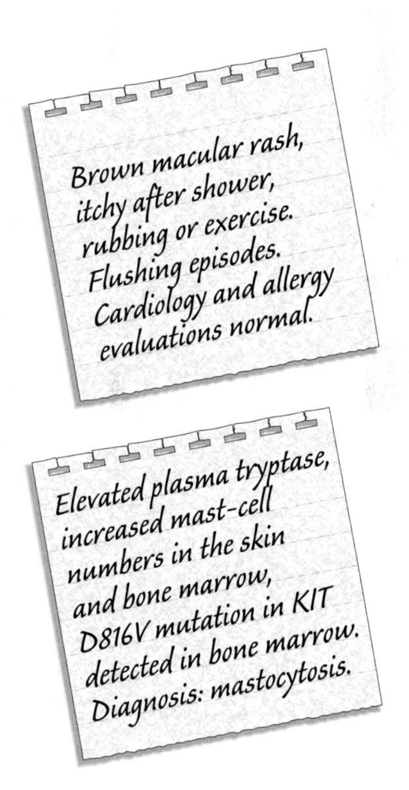

Brown macular rash, itchy after shower, rubbing or exercise. Flushing episodes. Cardiology and allergy evaluations normal.
Elevated plasma tryptase, increased mast-cell numbers in the skin and bone marrow, D816V mutation in KIT detected in bone marrow. Diagnosis: mastocytosis.

Giles's blood pressure and his heart rate and rhythm were normal. His nasal turbinates were not swollen and his lungs were clear on auscultation, with no wheezing or rales (crackles). The abdomen was soft, and slightly tender to deep palpation over the epigastric area with no rebound. There was no liver or spleen enlargement. Examination of the skin revealed mild flushing of the face. Multiple small hyperpigmented macular (flat) lesions (urticaria pigmentosa; Fig. 20.3) covered his trunk and extremities, except for areas exposed to the sun—hands, forearms, and face. These lesions become hives (urticated) within 2 minutes of rubbing with a tongue depressor, a physical finding known as Darier's sign.

Laboratory evaluations showed white blood cell, hemoglobin, and platelet counts to be within the normal range. There was mild eosinophilia (500 cells μl^{-1}). Total serum IgE was 10 IU ml^{-1}, within the normal range. No allergen-specific IgE antibodies against foods or inhalant allergens were detected. Levels of IgE specific for honeybee and wasp venoms, measured by a fluorenzyme immunoassay (see Fig. 12.5) were slightly raised. The total tryptase level (see Case 16) was elevated, at 53 ng ml^{-1} (normal: less than 12 ng ml^{-1}). A 24-hour measurement of the serotonin metabolite 5-hydroxyindoleacetic acid in the urine was within the normal range.

A skin biopsy showed approximately 20 times more mast cells in the upper dermis than normal, primarily surrounding blood vessels. There was some increase in the pigment melanin but no other inflammatory infiltrate. A bone marrow biopsy and aspiration showed a cellularly normal marrow with normal maturation of erythroid, myeloid, and lymphoid elements. Tryptase staining of the biopsy revealed multiple aggregates of mast cells located in paratrabecular and perivascular locations (Fig. 20.4). There was also an increase in mast cells in an interstitial pattern. These mast cells were elongated in shape and appeared to be hypogranulated, and had eccentric nuclei. Staining of serial sections for the interleukin-2 receptor α chain (CD25), not normally expressed by mast cells, showed tryptase-positive mast cells expressing this marker. The bone marrow aspirate tested positive for the mutation D816V in the gene *KIT* (aspartic acid to valine at position 816 in the protein), indicating the presence of a clone of progenitor cells that had undergone a somatic mutation.

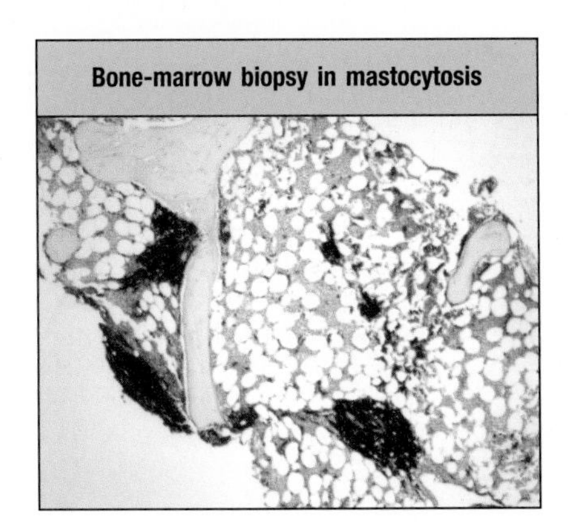

Fig. 20.4 Bone-marrow biopsy in mastocytosis. Bone-marrow biopsy showing multifocal dense mast-cell infiltrates in bone marrow highlighted by brown tryptase staining.

From the results of the bone marrow biopsy, Giles was diagnosed with indolent (non-progressive) systemic mastocytosis. He was started on the type 1 histamine receptor blocker fexofenadine (180 mg once daily) and the type 2 histamine receptor inhibitor ranitidine (150 mg twice daily), both taken orally, which greatly improved the symptoms of flushing, lightheadedness, and abdominal cramping. Giles was prescribed two epinephrine (adrenaline) autoinjectors and taught how to use them in case of the onset of a fainting episode. In addition, he was started on venom immunotherapy injections with a plan to keep him on these injections indefinitely. He was advised to have a bone densitometry screen, because systemic mastocytosis can predispose to accelerated osteoporosis. He was also counseled about potential triggers for mast-cell degranulation, such as temperature changes, stress, exercise, alcohol, and some medications, including nonsteroidal anti-inflammatory drugs (NSAIDs), opioids, and muscle relaxants. Because Giles had been able to take aspirin and the NSAID ibuprofen before the onset of his symptoms, he was told he could continue to take them. If Giles were to require surgery under general anesthesia, the anesthesiologist and the surgeon should be told of the diagnosis of mastocytosis, and premedication with a type 1 antihistamine (such as diphenhydramine) should be considered before surgery. Giles was told to make periodic follow-up visits to check his tryptase levels and differential white blood cell count.

Mastocytosis.

Mastocytosis, a primary disorder of mast-cell proliferation, can occur in children and adults. As in Giles's case, a helpful clue to the diagnosis is the presence of skin lesions of the type called urticaria pigmentosa (see Fig. 20.3). In contrast to typical urticarial wheals (hives; see Fig. 11.1), which are temporary and intensely itchy, urticaria pigmentosa lesions are permanent, hyperpigmented, and generally do not become hives unless exposed to a trigger such as heat or rubbing. One of the clues to diagnosis is a positive 'Darier's sign,' in which stroking the macule with a sharp object leads to urtication (the formation of hives). Mastocytosis in children is generally limited to the skin and resolves or improves by adolescence. Mastocytosis diagnosed in adults, in contrast, generally also involves internal organs (commonly the bone marrow) and is persistent. The most recent World Health Organization (WHO) classification of mastocytosis distinguishes seven clinical forms (Fig. 20.5). Of these, cutaneous and indolent systemic mastocytosis are the most common variants in children and adults, respectively. It is unusual for these disorders to progress to the aggressive forms of mastocytosis associated with decreased life expectancy.

Patients in all categories of mastocytosis can display symptoms caused by mast-cell degranulation and activation (Fig. 20.6). Symptoms commonly include episodic flushing, as experienced by Giles, accompanied by tachycardia (heart rate exceeding normal), abdominal cramping, nausea, vomiting, lightheadedness, feeling faint (presyncope), and fainting (syncope). Urticaria and angioedema occur only rarely during flares of mastocytosis resulting from mast-cell activation. The symptoms are caused by a diverse array of vasoactive mediators released by activated mast cells, such as histamine, prostaglandin D_2, and cysteinyl leukotrienes.

In most cases, the mast-cell accumulation in mastocytosis is clonal in nature. Mast cells in affected patients arise from a single abnormal hematopoietic progenitor carrying a mutation in *KIT*, and they display surface markers not usually found on mast cells, such as CD25. Thus, the disease can be regarded as a neoplastic disorder and is distinguished from increases in mast cells in

WHO classification of mastocytosis	
Cutaneous mastocytosis	Disease limited to skin with no bone marrow or internal organ involvement. Most commonly diagnosed in children within the first year of life. Usually improves or regresses by adolescence
Indolent systemic mastocytosis	Most common category in adults. May or may not present with skin lesions. Bone marrow involvement almost universal. Usually has a good prognosis with low risk of progression into an advanced category
Systemic mastocytosis with an associated hematologic non-mast-cell disorder	Approximately 10–20% of cases at diagnoses. Hematologic disease generally involves the myeloid lineage but may also be seen with lymphoproliferative disorders. Prognosis is poorer and depends on the associated hematologic disease
Aggressive systemic mastocytosis	Rare category characterized by end-organ dysfunction due to mast-cell infiltration. Examples include liver involvement with portal hypertension and ascites, extensive marrow involvement or splenomegaly with clinically significant cytopenias, gastrointestinal infiltration with malabsorption, diarrhea and weight loss, or large osteolytic lesions with pathologic fractures. Poor prognosis. Patients are candidates for cytoreductive therapy
Mast-cell leukemia	Rare category. Presence of immature mast cells > 10% in peripheral blood or > 20% in bone marrow aspirates. Poor prognosis
Mast-cell sarcoma	Rare category. Invasive solid mast-cell tumor
Extracutaneous mastocytoma	Rare. Benign, non-invasive solid mast-cell tumor

Fig. 20.5 World Health Organization (WHO) classification of mastocytosis.

response to a specific stimulus, which may be seen in a number of atopic, inflammatory, and neoplastic diseases.

Mast-cell activation in mastocytosis is generally due to non-IgE-mediated mechanisms such as direct activation by temperature changes, physical stimulation, emotional stress, or exercise (see Fig. 20.6), but can also occur unprovoked. Other known triggers include alcohol, bee or wasp stings, and certain medications such as opioids, NSAIDs, and muscle relaxants. IgE-mediated allergic diseases occur in patients with mastocytosis at a similar frequency to that in the general population; the allergic symptoms may, however, be more pronounced as a result of the greater numbers of mast cells in the tissues and the consequent increase in mediator release after FcεRI-triggered mast-cell activation. For example, in some studies, up to 10% of patients with systemic reactions induced by hymenopteran venom (see Case 16) had a clonal mast-cell disease as the underlying diagnosis. In contrast, food allergies and allergic asthma are not commonly associated with mastocytosis.

Serum tryptase level is a good screening marker for both mastocytosis and acute mast-cell activation events, including anaphylaxis (see Cases 1 and 16). Patients with mastocytosis generally have baseline tryptase levels higher than normal, correlating with the total body mast-cell burden; these levels can increase transiently after mast-cell degranulation. However, normal tryptase levels may be encountered in children with the cutaneous disease as well as in

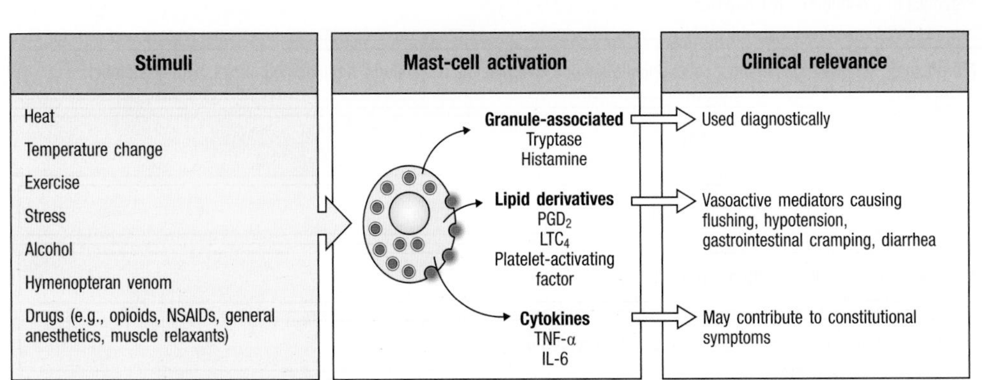

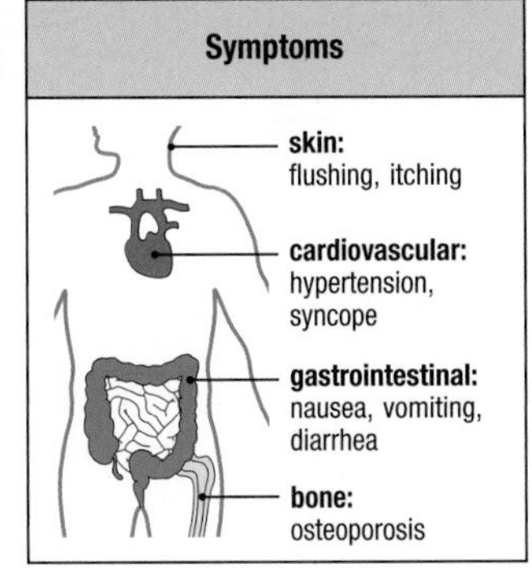

Fig. 20.6 Factors triggering mast-cell activation in mastocytosis, and its symptomatology. Stimuli that can trigger mastocytosis and the consequent mast-cell activation are shown in the first panel. Mast-cell activation results in the release of various vasoactive and other mediators (second panel) that cause symptoms in the skin and internal organs (third and fourth panels) and are also used for diagnosis.

adults with limited forms of disease, including monoclonal mast-cell expansion. Similarly, elevated tryptase levels may also be seen in patients with some myeloid neoplasms (myelodysplastic syndromes, acute leukemias, chronic eosinophilic leukemia), renal insufficiency, or sometimes without clear clinical significance (idiopathic).

A diagnosis of mastocytosis depends on histopathological features, and a bone marrow biopsy is required to establish the diagnosis of systemic mastocytosis (see Fig. 20.4). Because almost all adults with urticaria pigmentosa have bone marrow involvement, a bone marrow biopsy is recommended for this group to establish the diagnosis, to assess disease extent, and to determine the presence or absence of other non-mast-cell hematologic diseases in the marrow. In children, systemic involvement is rare, and bone marrow biopsy is reserved for those with unexplained persistent abnormalities in complete white blood cell count, enlargements of liver, spleen or lymph nodes, or persistently elevated tryptase levels greater than 20 ng ml^{-1}.

Mast-cell abnormalities in the bone marrow of patients with systemic mastocytosis include morphologic changes (such as cells with spindle shapes, cytoplasmic projections, or hypogranularity, occurring in clusters), aberrant cell-surface expression of CD25, and the presence of an activating mutation in *KIT*. The WHO criteria have to be met for the diagnosis to be made (Fig. 20.7). Those patients who do not fully meet the WHO criteria (having one or two minor criteria) are considered to have monoclonal mast-cell activation syndrome. These patients present with recurrent symptoms of mast-cell activation such as flushing, tachycardia, hypotensive syncope, and gastrointestinal cramping, but they may lack the skin lesions of urticaria pigmentosa and can have normal baseline tryptase levels. Hymenopteran stings are recognized to be a major trigger for a subset of these patients.

The most commonly reported mutation in the *KIT* gene in mastocytosis changes an essential aspartic acid (D) residue in the tyrosine kinase enzymatic pocket to valine (V) (D816V). This mutation results in ligand-independent

Fig. 20.7 WHO diagnostic criteria for systemic mastocytosis.

WHO diagnostic criteria for systemic mastocytosis	
Symptoms	
Major	Multifocal aggregates of 15 or more cells in bone marrow or other tissues other than skin (see Fig. 20.3)
Minor	Baseline serum tryptase level > 20 ng ml^{-1}
	Expression of CD25 by mast cells (detectable by immunohistochemistry or flow cytometry)
	Presence of a codon 816 KIT mutation
	Morphologic abnormalities in mast cells: spindle shapes, cytoplasmic projections, multilobated nuclei, hypogranulation
Diagnosis	
	One major + one minor: systemic mastocytosis
	No major but three minor: systemic mastocytosis
	One or two minor: monoclonal mast-cell activation syndrome

autoactivation of the receptor (see Fig. 20.1). The Kit-activating D816V mutation has been detected in mast cells in skin lesions in approximately 40% of children with cutaneous mastocytosis and in more than 95% of adults with systemic mastocytosis, regardless of the category. Therefore other, yet to be identified, mutations are likely to contribute to the disease phenotype.

Treatment of mastocytosis is primarily aimed at controlling or preventing the symptoms caused by mast-cell mediator release. Commonly used medications include blockers of the type 1 and type 2 histamine receptors, cromolyn sodium (a compound that inhibits mast-cell activation), and leukotriene inhibitors. Systemic glucocorticoids may be required in rare patients with refractory symptoms. At least two self-injectable epinephrine devices should be prescribed to each patient regardless of their history of anaphylaxis. Patients with advanced disease categories such as aggressive systemic mastocytosis and mast cell leukemia are candidates for treatments that reduce mast-cell numbers (cytoreduction). The most commonly used cytoreductive agents are interferon-α and cladribine (2-chlorodeoxyadenosine). These drugs lower the numbers of bone marrow and tissue mast cells in some patients but do not eliminate the disease. Tyrosine kinase inhibitors, including imatinib, have been investigated for the treatment of mastocytosis, but although imatinib strongly inhibits wild-type Kit, it is a very poor inhibitor of Kit with the D816V mutation and is therefore not an appropriate drug for most patients with mastocytosis.

Questions.

1 *Compare and contrast Giles's symptoms with those of idiopathic anaphylaxis.*

2 *Why is cytoreductive therapy not generally considered for patients with cutaneous or indolent systemic mastocytosis?*

3 *Why is imatinib not a good drug for most patients with mastocytosis?*

4 *What potentially deleterious conditions should be watched for in patients with systemic mastocytosis?*

5 *Why do hymenopteran stings pose a special risk for patients with mastocytosis?*

6 *Why was Giles prescribed epinephrine injectors?*

Answers

Case 1

Answer 1

John's hoarseness resulted from angioedema of the vocal cords. His wheezing was due to forced expiration of air through bronchi that had become constricted. In this case, constriction resulted from the release by activated mast cells of histamine and leukotrienes that caused the smooth muscles of the bronchial tubes to constrict.

Answer 2

John's parents were instructed to avoid feeding him any food containing peanuts and to read the labels of packaged foods scrupulously to avoid anything containing peanuts. They were advised to inquire in restaurants about food containing peanuts. Because green peas, also a legume, contain an antigen that cross-reacts with peanuts and might also incite an anaphylactic reaction, peas were withdrawn from John's diet. A Medi-Alert bracelet, indicating his anaphylactic reaction to peanuts, was ordered for John. The parents were also given an Epi-Pen syringe pre-filled with epinephrine to keep at home or while traveling, in case John developed another anaphylactic reaction.

Answer 3

Epinephrine acts at β_2-adrenergic receptors in smooth muscle surrounding blood vessels and bronchi. It has opposing effects on the two types of muscle. It contracts the muscle surrounding the small blood vessels, thereby constricting them, stopping vascular leakage, and raising the blood pressure. It relaxes that of the bronchi, making breathing easier.

Answer 4

Histamine and tryptase are released by activated mast cells; high levels in the blood indicate the massive release from the mast cells that occurs during an anaphylactic reaction.

Answer 5

Immediately after a systemic anaphylactic reaction the patient is unresponsive in a skin test owing to the massive depletion of mast-cell granules and failure of the blood vessels to respond to mediators. This is called tachyphylaxis and lasts for 72–96 hours after the anaphylactic reaction. For this reason, John had to come back to the Allergy Clinic a few days later for his tests.

Answer 6

Increased incidence of peanut allergy has been related to the increasing topical use of peanut-oil-based creams to treat dry skin in infants. Although peanut oil is not the culpable allergen, it is often contaminated by allergenic peanut proteins.

Answer 7

It has been shown that depletion of IgE by the administration of a humanized mouse anti-human IgE antibody that binds circulating IgE, but not mast-cell-bound IgE, results in protection from peanut anaphylaxis. This therapy works because it results in the eventual depletion of mast-cell-bound IgE, which is in equilibrium with serum IgE. It is safe because the anti-IgE antibody does not trigger mast-cell activation.

Case 2

Answer 1

During inspiration, the negative pressure on the airways causes their diameter to increase, allowing an inflow of air. During expiration, the positive expiratory pressure tends to narrow the airways. This narrowing is exaggerated when the airway is inflamed and bronchial smooth muscle is constricted, as in asthma. This causes air to be trapped in the lungs, with an increase in residual lung volume at the end of expiration. Breathing at high residual lung volume means more work for the muscles and increased expenditure of energy; this results in the sensation of tightness in the chest. The high residual lung volume is also the cause of the hyperinflated chest observed on the chest radiograph. The peribronchial inflammation in asthma causes bronchial marking around the airways.

Answer 2

Chronic allergic asthma is not simply due to constriction of the smooth muscles that surround the airway: it is largely due to the inflammatory reaction in the airway, which consists of cellular infiltration, increased secretion of mucus, and swelling of the bronchial tissues. This explains the failure of bronchodilators, which dilate smooth muscles, to maintain an open airway and their failure to completely reverse the decreased air flow during Frank's acute attacks. Steroids are therefore given to combat the inflammatory reaction of the late-phase response.

Answer 3

Allergic individuals have a tendency to respond to allergens with an immune response skewed to the production of T_H2 cells rather than T_H1. The cells produce the interleukins IL-4 and IL-13, cytokines that induce IgE production in humans. T_H2 cells also make IL-5, which is essential for eosinophil maturation. Furthermore, activated T cells and bronchial epithelial cells secrete CCL11 (formerly known as eotaxin), which attracts eosinophils in the airways. The production of IL-4 and IL-5 by T_H2 cells responding to allergens in atopic individuals explains the frequent association of IgE antibody response and eosinophilia in these patients.

Answer 4

IgE-mediated hypersensitivity to an allergen is tested for by injecting a small amount of the allergen intradermally. In allergic individuals, this is followed within 10–20 minutes by a wheal-and-flare reaction at the site of injection (see Fig. 2.5), which subsides within an hour. The wheal-and-flare reaction is due mainly to the release of histamine by mast cells in the skin. This increases the permeability of blood vessels and the leakage of their contents into the tissues, resulting in the swollen wheal; dilation of the fine blood vessels around the area produces the diffuse red 'flare' seen around the wheal. This reaction is almost completely inhibited by antagonists of the histamine type 1 receptor, the major histamine receptor expressed in the skin.

Answer 5

The recurrence of the redness and swelling at the site of previous immediate allergic reactions represents the late-phase response characterized by a cellular infiltrate.

Answer 6
Nonsteroidal anti-inflammatory drugs (NSAIDs) such as aspirin and ibuprofen can induce wheezing in certain patients (see Case 19). This is classically seen in patients with Sampter's triad: asthma, nasal polyps, and NSAID sensitivity. NSAIDs inhibit the enzyme cyclooxygenase (COX). Normally, the actions of COX lead to the synthesis of prostaglandins from arachidonic acid. COX inhibition leads to shunting of the arachidonic acid precursor away from prostaglandin synthesis and into the leukotriene synthesis pathway (see Fig. 2.7). The increased leukotriene biosynthesis leads to bronchial smooth muscle constriction and cell proliferation, plasma leakage, mucus hypersecretion, and eosinophil migration, culminating in symptoms of wheezing and asthma exacerbation. Leukotriene E4 levels can be measured in the urine. In patients with aspirin sensitivity, E4 levels are higher at baseline and rise an additional fivefold after aspirin ingestion before returning to baseline as the aspirin-induced wheezing resolves.

Answer 7
Repeated administration of relatively high doses of allergen by subcutaneous injection is thought to favor antigen presentation by antigen-presenting cells that produce IL-12. This results in the induction of T_H1 cells rather than T_H2 cells. The presence of T_H1 cells tends to lead to an IgG antibody response rather than an IgE response because the T_H1 cells produce IFN-γ, which prevents further isotype switching to IgE. The IgG antibody competes with the IgE antibody for antigen. Furthermore, IgG bound to allergen inhibits mast-cell activation (via FcεRI) and B-cell activation (via surface immunoglobulin) by allergen because of inhibitory signals delivered subsequent to the binding of Fcγ receptors on these cells. This is thought to be one mechanism damping down the allergic response. Another is no further boosting of IgE production because IL-4 and IL-13 are not secreted. Existing IgE levels themselves may not fall by much, because IFN-γ does not affect B cells that have already switched to IgE production.

Answer 8
Most human allergy is caused by a limited number of inhaled protein allergens that elicit a T_H2 response in genetically predisposed individuals. These allergens are relatively small, highly soluble protein molecules that are presented to the immune system by the mucosal route at very low doses. It has been estimated that the maximum exposure to ragweed pollen allergens is less than 1 μg per year. It seems that transmucosal presentation of very low doses of allergens favors the activation of IL-4-producing T_H2 cells and is particularly efficient at inducing IgE responses. The dominant antigen-presenting cell type in the respiratory mucosa expresses high levels of co-stimulatory B7.2 molecules. Expression of B7.2 on antigen-presenting cells is thought to favor the development of T_H2 cells. In contrast, injection of antigen subcutaneously in large doses, as occurs on vaccination, results in antigen uptake in the local lymph nodes by a variety of antigen-presenting cells and favors the development of T_H1 cells, which inhibit antibody switching to IgE.

Case 3

Answer 1
The daytime pollen exposure causes the early-phase reaction, accounting for Charlie's immediate symptoms, followed 8–12 hours later by leukocyte recruitment and symptoms of the late-phase reaction.

Answer 2
The inflammation caused by allergic rhinitis can lead to obstruction of the

sinuses, resulting in the trapping of mucus. The stagnant mucus provides an excellent medium for bacterial growth.

Answer 3

Although allergy shots have proved very effective in reducing allergen-induced symptoms in people with allergic rhinitis, there are significant pitfalls associated with this approach. IgE production continues in patients receiving shots and in some cases the injection of allergen can provoke immediate hypersensitivity reactions. These can manifest as local reactions with hives or swelling at the injection site or, more seriously, as systemic reactions that can progress to anaphylaxis. For this reason, shots should only be given in a physician's office, and patients are generally monitored for 30 minutes after their injections. All patients receiving allergen immunotherapy are trained in the use of epinephrine autoinjectors so that they can manage allergic reactions that might occur after they go home. Patients with poorly controlled asthma are at particular risk of adverse reactions and may not be good candidates for immunotherapy.

Case 4

Answer 1

Although histamine is a very important mediator of allergic symptoms in seasonal allergic conjunctivitis, several other mast-cell products are also present and active (see Fig. 4.2). These include proteases, cytokines, chemokines, and lipid mediators which can act to increase blood flow (redness), cause plasma extravasation (edema), recruit inflammatory cells, and cause pruritis. One commonly used strategy to reduce exposure to all of these mediators is to administer mast-cell stabilizing compounds which serve to block allergen-triggered mast-cell activation.

Answer 2

Patients with vernal keratoconjunctivitis typically have more severe ocular involvement with intense itching and photophobia. A thick, ropy, ocular discharge can be seen. The conjunctivae have a cobblestone appearance. Careful ocular exam can reveal the presence of Horner–Trantas dots, which are collections of eosinophils and epithelial cells.

Answer 3

Upon initial exposure to an antigen, macrophages and dendritic cells present a peptide fragment of the antigen to a CD4 helper T cell (T_H2 cell). After activation, T_H2 cells express CD40 ligand on their surface and secrete the cytokine IL-4. These signals act together to induce the process of immunoglobulin class switch recombination to IgE (see Case 3) in B cells. IgE antibodies bind to FcεRI receptors on the surface of mast cells with very high affinity. Upon exposure to the allergen, cross-linking of IgE antibodies bound to mast cells via FcεRI results in mast-cell degranulation and the release of allergic mediators.

Answer 4

Allergen skin testing has limited value in situations where mast cells or their mediators are blocked (thereby inhibiting the cutaneous wheal-and-flare response) or where nonspecific mast-cell activation is present (as in mastocytosis syndromes). Usually, patients are advised to stop all antihistamine medications for 2 weeks before allergy evaluation. Some individuals are unable to comply with this recommendation. When skin testing is performed, histamine is applied as a positive control. If pharmacologic histamine blockade is still present, this test is negative and alternative methods of evaluation must be considered. In patients with mast-cell activation, the negative control (normal saline) gives a wheal-and-flare response, indicating that the mere physical stimulation of the skin prick is sufficient to induce mediator release. In this situation, the meaning of any other 'positive' reactions becomes suspect. Blood tests for allergen-specific IgE provide a very useful alternative in these settings, because they are not affected by the function of mast cells or their mediators.

Case 5

Answer 1

Corticosteroids bind to steroid receptors in inflammatory cells such as T cells and eosinophils. The steroid:receptor complex is translocated into the nucleus, where it can control gene expression, including the expression of cytokine genes, by binding to control elements in the DNA. In addition, corticosteroids increase the synthesis of the inhibitor of the transcription factor NFκB, which controls the expression of multiple cytokine genes. One effect is to inhibit the synthesis of cytokines and the release of preformed mediators and arachidonic acid metabolites. Although topical steroids are very effective, excessive or prolonged use of powerful steroids can lead to local skin atrophy.

Answer 2

The immunosuppressant cyclosporin A acts primarily on T cells and interferes with the transcription of cytokine genes. The drug binds to an intracellular protein, cyclophilin, and this complex in turn inhibits calcineurin, which normally dephosphorylates nuclear factor of activated T cells (NFAT), a major cytokine gene transcription factor. FK506, or tacrolimus, is another immunosuppressant with a spectrum of activity similar to that of cyclosporin. Tacrolimus binds to the cytoplasmic protein FK506-binding protein, and this complex also inhibits calcineurin. Tacrolimus has a smaller molecular size and higher potency than cyclosporin A and, perhaps because of these features, it seems to be effective as a topical formulation.

Answer 3

Patients with atopic dermatitis have defective local innate cell-mediated immunity, which is required for the control of herpesvirus and vaccinia virus infections: T_H2 cytokine expression in the affected skin inhibits the production of antimicrobial peptides by keratinocytes. Cell-mediated adaptive immune responses involve T_H1 CD4 cells and CD8 cytotoxic cells; patients with atopic dermatitis have selective activation of T_H2 rather than T_H1 cells, as shown by their reduced delayed-type hypersensitivity skin reactions. They also have decreased numbers and function of CD8 cytotoxic T cells. Furthermore, monocytes from patients with atopic dermatitis secrete increased amounts of IL-10 and prostaglandin E_2 (PGE_2). Both IL-10 and PGE_2 inhibit the production of the T_H1 cytokine IFN-γ, and IL-10 also inhibits T-cell-mediated reactions.

Answer 4

Scratching causes tissue damage that stimulates the keratinocytes to secrete cytokines and chemokines (IL-1, IL-6, CXCL8, GM-CSF, and TNF-α). IL-1 and TNF-α induce the expression of adhesion molecules such as E-selectin, ICAM-1, and VCAM-1 on endothelial cells, which attract lymphocytes, macrophages, and eosinophils into the skin. These infiltrating cells secrete cytokines and inflammatory mediators that perpetuate keratinocyte activation and cutaneous inflammation.

Answer 5

The skin of more than 90% of patients with atopic dermatitis is colonized by *Staphylococcus aureus*. Recent studies suggest that *S. aureus* can exacerbate or maintain skin inflammation in atopic dermatitis by secreting a group of toxins known as superantigens, which cause polyclonal stimulation of T cells and macrophages. T cells from patients with atopic dermatitis preferentially express T-cell receptor β chains $V_\beta 3$, 8, and 12, which can be stimulated by staphylococcal superantigens, resulting in T-cell proliferation and increased IL-5 production. Staphylococcal superantigens can also induce expression of the skin homing receptor (CLA) in T cells, which is mediated by IL-12. In addition, nearly half of patients with atopic dermatitis produce IgE directed against staphylococcal superantigens, particularly SEA, SEB, and toxic shock syndrome toxin-1 (TSST-1). Basophils from patients with atopic dermatitis who produce antitoxin IgE release histamine on exposure to the relevant toxin. These findings suggest that local production of staphylococcal exotoxins at the skin surface could cause IgE-mediated histamine release and thereby trigger the itch–scratch cycle that exacerbates the eczema.

Answer 6

A mouse model of atopic dermatitis suggests that sensitization directly through the skin can result in allergen-induced asthma. In this model, patch application of allergen to the shaved skin of a normal mouse results in an eczematous dermatitis and subsequent allergen-specific airway hypersensitivity such that exposure to allergen by inhalation causes airway hyperresponsiveness typical of the asthmatic state.

There is epidemiologic evidence that sensitization of infants to food allergens through the skin may predispose to food allergy, and that unlike oral exposure to food allergens it does not induce tolerance, but rather results in IgE antibody formation that can cause anaphylaxis.

Case 6

Answer 1

Pentadecacatechol can be transferred from the initial point of contact to other areas of the skin by the fingernails after scratching the itchy lesion at the primary site of hapten introduction. This is why it is essential to cut the fingernails short and thoroughly wash off the skin and scalp to remove the chemical and prevent further spread.

Answer 2

The half-life of some of the proteins haptenated by pentadecacatechol can be quite long. CD4 memory T cells will continue to be activated as long as the haptenated peptides are being generated. In Paul's case this went beyond the third week after contact with poison ivy.

Answer 3

Once an individual has been sensitized, the reaction often becomes worse with each exposure, as each reexposure not only produces the hypersensitivity reaction but generates more effector and memory T cells. Memory T cells that mediate delayed hypersensitivity reactions, such as contact sensitivity to poison ivy and the tuberculin test, can persist for most of the life of the individual.

Answer 4

You could perform a patch test. In this test a patch of material impregnated with the hapten is applied to the skin under seal for 48 hours. The area is then examined for redness, swelling, and vesicle formation. Alternatively, peripheral blood mononuclear cells can be incubated with the hapten and T-cell proliferation assessed 6–9 days later.

Answer 5

The risk of Brian's developing poison ivy sensitivity is at least as high as that for a normal child. This is because antibody plays no discernible role in the genesis of delayed hypersensitivity reactions, and T-cell function is normal in X-linked agammaglobulinemia. In fact, clinical observations suggest that boys with X-linked agammaglobulinemia may develop more severe forms of poison ivy sensitivity. It has been suggested that, in the absence of antibody, more hapten is available for conjugation with self proteins and that, in the absence of antigen presentation by β cells, the T-cell response is skewed more towards T_H1 cells.

Answer 6

The artificially induced tuberculin reaction is a good model of a delayed hypersensitivity reaction. This skin test detects infection with the bacterium *Mycobacterium tuberculosis*, or previous immunization against tuberculosis with the live attenuated vaccine BCG. Small amounts of tuberculin, a protein derived from *M. tuberculosis*, are injected subcutaneously; a day or two later, a sensitized person develops a small, red, raised area of skin at the site of injection. In countries where BCG is administered routinely to babies, the tuberculin test can be used to test for T-cell function. This is because antigen-specific memory T cells are long-lived, and the sensitivity to tuberculin will persist throughout life. In the USA, children are not immunized with BCG. However, they all receive a full course of diphtheria and tetanus vaccines, which in each case contain purified protein toxoids as the antigen. Contact sensitivity to these two antigens can be used to test T-cell function. Alternatively, antigen derived from the yeast-like fungus *Candida albicans*, which is a normal inhabitant of the body flora, can be used to induce a delayed hypersensitivity reaction in the skin.

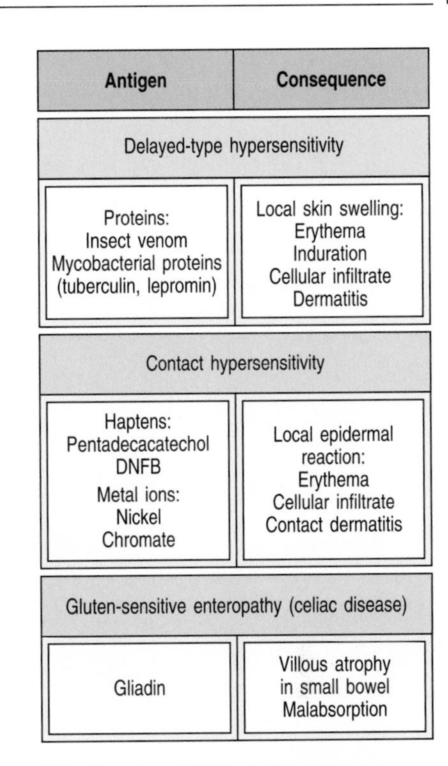

Antigen
Consequence
Delayed-type hypersensitivity
Proteins:
Insect venom
Mycobacterial proteins
(tuberculin, lepromin)
Local skin swelling:
Erythema
Induration
Cellular infiltrate
Dermatitis
Contact hypersensitivity
Haptens:
Pentadecacatechol
DNFB
Metal ions:
Nickel
Chromate
Local epidermal
reaction:
Erythema
Cellular infiltrate
Contact dermatitis
Gluten-sensitive enteropathy (celiac disease)
Gliadin
Villous atrophy
in small bowel
Malabsorption

Fig. A6.7 Some type IV hypersensitivity reactions. Depending on the source of antigen and its route of introduction, these clinical conditions have different names and consequences. DNFB, dinitrofluorobenzene.

Answer 7

Some of the commoner environmental causes of delayed hypersensitivity reactions are insect bites or stings, which introduce insect venom proteins under the skin, and skin contact with chemicals in the leaves of some plants, or with metals such as nickel, beryllium, and chromium (Fig. A6.7). Nickel sensitivity is quite common and often occurs at the site of contact with nickel-containing jewelry. Contact sensitivity to beryllium has been well documented in factory workers engaged in manufacturing fluorescent light bulbs. Celiac disease is a type of delayed sensitivity reaction seen in people who are allergic to the protein gliadin, a constituent of wheat grains and flour. Patients with celiac disease therefore have to avoid all food products containing wheat flour.

Case 7

Answer 1

Antibodies against other plasma or platelet proteins can trigger anaphylaxis. For example, anti-haptoglobin antibodies can be present in individuals with haptoglobin deficiency. This is a common cause of transfusion reactions in Japan. Hemolysis due to anti-ABO antibodies, the cause of reactions in unmatched red-cell transfusions, can also occur after platelet transfusion. Type O blood, which is sometimes used in emergencies when the recipient's blood type is not known, can have high titers of anti-A/B antibodies in the plasma, which can lead to intravascular hemolysis when donor platelets are transfused into a type A, AB, or B recipient. Platelet products are typically screened for high titers of anti-A/B antibodies; in Nathan's case, ABO-matched platelets were used.

Answer 2

If Nathan needed a red-cell transfusion, red cells can be washed repeatedly to ensure that plasma has been removed. Thawed frozen red cells can also be used because, before freezing, red cells are extensively washed and placed in storage solutions that do not contain plasma. Platelets can also be washed to remove plasma, but in Nathan's case, which involved a severe reaction, it would be prudent to obtain platelets from IgA-deficient donors. If needed, plasma would have to be derived from IgA-deficient donors.

Answer 3

IVIG preparations contain mainly IgG, but small amounts of IgA are present. Given that Nathan has a history of anaphylactic reaction to IgA, it would not be advisable to use IVIG in Nathan as a first-line therapy for ITP, given the availability of other treatment options (steroids or other immunomodulators). Amounts of IgA vary between manufacturers, however, and in the case of a patient with known IgA deficiency but no previous history of anaphylaxis on transfusion, you could consider selecting a product containing low levels of IgA and performing the initial infusion slowly, observing the patient closely for any signs of reaction.

Answer 4

At present such screening would not be efficient or cost-effective in preventing transfusion reactions. The presence of anti-IgA antibodies does not necessarily indicate that an anaphylactic reaction will occur upon exposure to IgA. In addition, anti-IgA antibody levels can vary over time, so testing would be best left until the time a transfusion is needed. Last but not least, completely reliable assays for anti-IgA antibodies are not yet available.

Answer 5

Bacterial contamination of platelet products can lead to sepsis, which can share symptoms with an anaphylactic reaction. One distinguishing symptom is fever, which would be seen with sepsis. Platelets are stored at room temperature and can be contaminated with donor skin bacteria at the time of needle insertion for collection. To avoid contamination, the first few milliliters of collected blood are diverted and discarded. In addition, assays are performed on platelets in the blood bank to detect bacterial growth.

Case 8

Answer 1

The symptoms were caused by the activation of complement-generated C3a, which releases histamine from mast cells and causes hives. The swelling around the mouth and eyelids is a form of angioedema. There is a more complete discussion of the role of the complement and the kinin systems in the pathogenesis of angioedema in Case 10.

Answer 2

Gregory almost certainly had developed vasculitis in the small blood vessels of his brain, and this compromised oxygen delivery to his brain.

Answer 3

He had red cells and albumin in his urine, which indicated an inflammation of the small blood vessels in his kidney glomeruli. He also developed purpura in his feet and ankles. Purpura (which is the Latin word for purple) indicates hemorrhage from small blood vessels in the skin that are inflamed and have become plugged with clots. A skin biopsy of one lesion showed the deposition of IgG and C3 around the small blood vessels, suggesting that an immune reaction was taking place.

Answer 4

You would expect to see massive follicular hyperplasia, polyclonal B-cell activation, and many mature plasma cells in the medulla. The massive B-cell activation in the lymph nodes leads to an overflow of plasma cells from the medulla of the nodes into the efferent lymph. It is otherwise very, very rare to find plasma cells in the blood, as were found in Gregory's blood. They find their way to the bloodstream via the thoracic duct. The enlargement of the spleen was almost certainly due to hyperplasia of the white pulp. Some plasma cells probably enter the blood from the hyperplastic follicles in the spleen.

Answer 5

The acute-phase reaction is caused by interleukin (IL)-1 and to a greater extent by IL-6, which are released from monocytes that have been activated by the uptake of immune complexes. The acute-phase response consists of marked changes in protein synthesis by the liver. The synthesis of albumin drops sharply, as does the synthesis of transferrin. The synthesis of fibrinogen, C-reactive protein, amyloid A, and several glycoproteins is rapidly upregulated. The precise advantage to the host of the acute-phase reaction is not well understood, but it is presumably a part of innate immunity, which aids host resistance to pathogens before the adaptive immune system becomes engaged.

Answer 6

His serum C1q level was decreased. This almost always indicates complement consumption by immune complexes via the classical pathway. (In hereditary angioedema (see Case 10) the C1q level is normal; in this disease, complement is activated because of a defect in an inhibitor and not by the formation of immune complexes.) The level of C3 in Gregory's serum was also lowered, a further indication of complement consumption.

Answer 7

No! The skin test is positive when there are IgE antibodies bound to the mast cells in the skin. Gregory did not have IgE antibodies against penicillin, as confirmed by the negative fluorenzyme immunoassay. Serum sickness is caused by complement-fixing IgG antibodies.

Case 9

Answer 1

In Foxp3-deficient mice, infusion of relatively small numbers of T_{reg} cells controls the disease symptoms. So it is likely that, in humans also, a small number of natural T_{reg} cells is sufficient for immune regulation, and Billy can make sufficient Treg $_{cells}$ that derive from his sister's bone marrow stem cells.

Answer 2

The 'conditioning' leading up to a bone marrow transplant comprises treatment with cytotoxic drugs that kill all rapidly dividing cells, including the CD4 effector T cells responsible for the uncontrolled inflammatory response in IPEX. As these cells are destroyed, the autoimmune response that they have produced will be dampened. Immunosuppressant drugs have the same effect in reducing the activation and proliferation of T cells, and this is why they are used to treat IPEX and other autoimmune disorders.

Answer 3
Because IPEX patients are unable to downregulate the immune activation triggered by infections, their disease frequently flares up on exposure to pathogens and even after vaccination. IVIG is useful in forestalling infections or ameliorating their impact when they occur. Vaccination is contraindicated in IPEX patients because of the risk of disease flare-up.

Answer 4
Venous infusion of immunocompetent purified naive CD4 T cells into mice with severe immunodeficiency (for example, *scid* mice or Rag-deficient mice) results in colitis. Infusion of CD4 CD25 T_{reg} cells into these recipients can both prevent and reverse the colitis. This suggests that CD4 CD25 T_{reg} cells may have therapeutic potential in human autoimmune diseases.

Answer 5
Patients with IL-2Rα deficiency present a clinical picture with similarities to IPEX. IL-2 signaling is essential for the maintenance of T_{reg} cells. Therefore, T_{reg} activity is deficient in the absence of IL-2, IL-2Rα (CD25), or the transcription factor STAT5, which is important for transducing the IL-2 signal. This may explain the similarity of the symptoms of IL-2Rα deficiency to those of IPEX.

Case 10

Answer 1
Histamine release on complement activation is caused by C3a (the small cleavage fragment of C3), and the main chemokine is C5a (the small cleavage fragment of C5). These are both generated by the C3/C5 convertase, which in the classical pathway is formed from C4b and C2a. In HAE, C4b and C2a are both generated free in plasma. C4b is rapidly inactivated if it does not bind immediately to a cell surface; for that reason, and because the concentrations of C4b and C2a are relatively low, no C3/C5 convertase is formed, C3 and C5 are not cleaved, and C3a and C5a are not generated.

The edema in HAE is caused not by the potent inflammatory mediators of the late events in complement activation, but by C2b generated during the early events, and by bradykinin generated through the uninhibited activation of the kinin system.

Answer 2
The only other complement component that should be decreased is C2, which is also cleaved by C1. C1 plays no part in the alternative pathway of complement activation, so complement activation by the alternative pathway is not affected. The terminal components are not affected either. The unregulated activation of the early complement components does not lead to the formation of the C3/C5 convertase (see Question 1), so the terminal components are not abnormally activated. The depletion of the early components of the classical pathway does not affect the response to the normal activation of complement by bound antibody because the amplification of the response through the alternative pathway compensates for the deficiency in C4 and C2.

Answer 3
This is not hard to explain; as we have already remarked, the alternative pathway of complement activation is intact and thus, although the classical pathway is affected by deficiencies in C2 and C4, these are compensated for by the potent amplification step from the alternative pathway.

Answer 4

Stanozolol is a well-known anabolic androgen that has been used illegally by Olympic competitors. For unknown reasons, anabolic androgens suppress the symptoms of HAE, and that is why stanozolol was prescribed to Richard. Patients, especially females, do not like to take these compounds because they cause weight gain, acne, and sometimes amenorrhea. Preparations of purified C1INH are now available, and intravenous injection of C1INH prepared from human donors is safe and very effective in halting the symptoms of the disease.

Answer 5

In practice, you would administer epinephrine immediately in any case, because most such emergencies are due to anaphylactic reactions and because epinephrine is a harmless drug. If the laryngeal edema is anaphylactic, it will respond to the epinephrine. If it is due to hereditary angioedema, it will not. Anaphylactic edema is also likely to be accompanied by urticaria and itching, and the patient may have been exposed to a known allergen. Most patients know if they are allergic or have a hereditary disease, and they should be asked whether they have had a similar problem before.

Answer 6

HAE does not skip generations: it is therefore likely that its effects are dominant. It clearly affects both males and females, so it cannot be sex-linked. If the gene has a dominant phenotype, and Richard's two children are normal, then it follows they cannot have inherited the defective gene from their father, and their children cannot inherit the disease from them.

Richard has inherited his abnormal *C1INH* gene from his mother. Because he has a normal *C1INH* gene from his father, you might expect that he would have 50% of the normal level of C1INH. However, the tests performed by his immunologist revealed 16% of the normal level. In general, functional C1INH tests in HAE patients reveal between 5% and 30% of normal activity. How could this be explained? There are two possibilities: decreased synthesis (that is, less than 50% synthesis from only one gene); or increased consumption of C1INH as a result of increased C1 activation. Both explanations have been shown to be correct. Patients with HAE synthesize about 37–40% of the normal amount of C1INH, and C1INH catabolism is 50% greater than in normal controls.

Case 11

Answer 1

Fitzwilliam's positive blood test for peanut allergy indicates that he has B cells that produce IgE specific for peanut antigens. This peanut-specific IgE is bound to the surfaces of mast cells throughout the body. When Fitzwilliam ate the peanut sandwich as a youngster, peanut antigens entered the circulation through the gastrointestinal tract. When the circulating peanut antigen encountered peanut-specific IgE bound through FcεRI to mast cells in the dermis, this triggered the release of vasoactive mediators, resulting in urticaria. Interaction of the systemically absorbed and circulating peanut antigen with IgE on mast cells residing in the airway simultaneously caused airway edema and bronchospasm (wheezing). More recently, when Fitzwilliam petted the cat, he absorbed antigen through the skin where he touched the cat. In this case, mast cells residing in the skin were activated when the cat antigen bound to antigen-specific IgE that had been captured by FcεRI on the mast cells' surface. However, because the cat antigen was not absorbed into the systemic circulation, mast cells in other tissues were not activated.

Answer 2
Epinephrine is a catecholamine that acts on adrenergic receptors to stimulate peripheral vasoconstriction. Thus, it can reverse the vasodilation induced by histamine and other mast-cell mediators, effectively reversing the process that causes acute hives. In acute anaphylaxis, epinephrine can be life-saving because restoration of peripheral vascular tone and enhancement of cardiac contractility help maintain systemic blood pressure and prevent shock. However, epinephrine is only used in an emergency because it can induce significant tachycardia (elevation in the heart rate). In some susceptible individuals, systemically administered epinephrine can induce arrhythmias or coronary ischemia.

Answer 3
There are several possible reasons. First, the stratum corneum (the outermost keratinized layer of the epidermis) is much thicker on the palms and the soles, making it more difficult for antigens to penetrate into the dermis, where mast cells are located. Second, the vasodilation and edema that produce urticaria arise from blood vessels and lymphatic vessels. In palms and soles, these vessels are located deeper in the skin than on other areas of the body. Thus, because of the thick outer layers, processes occurring at that depth might not lead to obvious lesions at the surface. However, skin areas normally exposed to more pressure are more likely to develop hives, as pressure itself can directly activate mast cells.

Answer 4
Cold-induced urticaria is classified as one of the physical urticarias, in which physical stimulus (in this case cold) directly causes mast-cell degranulation and the release of mediators that cause urticaria. Cold air, cold water, evaporative cooling after getting out of the bath, and contact with cold surfaces can all be triggers for cold-induced urticaria. Testing is often done with an ice cube or cold pack applied to the skin for 10–15 minutes. The test is positive if a wheal arises at the site of cold exposure after removal of the ice cube.

Case 12

Answer 1
Gregor is among the 70–80% of egg-allergic children who tolerate foods containing well-baked egg. Egg white contains more than 20 different glycoproteins. Ovalbumin is the most abundant but is sensitive to thermal denaturation. Ovomucoid comprises only 10% of the egg protein but is the dominant allergen and is heat resistant. Individuals can develop IgE antibodies against linear or conformational epitopes of either of these antigens. Studies of children with egg allergy suggest that IgE antibodies from patients with persistent egg allergy recognize more linear epitopes, whereas those of children with transient egg allergy are more directed to conformational epitopes. Gregor was probably preferentially sensitized to the heat-labile and conformational egg epitopes. This is consistent with his negative ovomucoid-specific IgE, because the ovomucoid IgE-binding epitope is heat stable.

Answer 2
The seasonal influenza and yellow fever vaccines are grown in chicken eggs. They therefore contain trace amounts of egg proteins and should be administered with caution in egg-allergic patients. In the past, administration of influenza vaccines was contraindicated in egg-allergic children. However, substantial experience with the influenza vaccine emerged during the 2009–10 H1N1 pandemic, leading to a recommendation that most egg-allergic

patients can receive influenza vaccination. Egg-allergic patients can receive either a two-dose injection (10% followed by 90% of the dose) or a single-dose regimen, and are observed for 30 minutes afterwards. Many vaccine manufacturers list their ovalbumin content ranges on the package insert, and it is recommended to administer the vaccine with the lowest ovalbumin content available. Although several studies of small numbers of patients have demonstrated that the influenza vaccine can be administered safely, even in patients even with a history of anaphylaxis to egg, larger studies are needed before a definitive recommendation can be made about giving the influenza vaccine to this group of patients. The intranasal influenza vaccine is not recommended in egg-allergic patients, primarily because the risk of allergic reactions has not been studied.

Because the yellow fever vaccine is also grown in chicken eggs, it is contraindicated in patients with egg allergy. However, it can be administered by a multi-step desensitization protocol to egg-allergic patients who require this vaccination. The literature on allergic reactions to the yellow fever vaccine in egg-allergic patients is sparse, and recommendations may change in the future, as they did for the influenza vaccine.

The MMR vaccine is grown in chick embryo fibroblast cultures and contains negligible or no egg protein. It can therefore be safely administered to egg-allergic children without further testing.

Answer 3

No. Although eggs and chicken meat come from the same bird, the antigens causing the food allergy are different. Similarly, most patients with cow's milk allergy can tolerate beef, because the major allergens are distinct.

Answer 4

Patients outgrowing their food allergy exhibit decreasing levels of allergen-specific IgE detected by blood testing or skin-prick test. However, often these numbers do not fall to zero, and IgE is detectable even after the allergy has resolved. For egg, a specific IgE level of 0.6 IU ml^{-1} corresponds to a more than 90% negative predictive value for passing a food challenge, based on a large study of children with atopic dermatitis and food allergies. This indicates that factors other than IgE levels have important roles in determining whether a person will have an allergic reaction to a food.

Answer 5

Parents of children with food allergies often ask what they should do to prevent the development of food allergy in their next child. At present, there is no definitive answer, and this is an area of emerging research and active debate. There is no evidence that restricting maternal diet during pregnancy and breastfeeding alters the long-term development of allergic disease. In infants at high risk for atopy (that is, infants with at least one first-degree relative with allergic disease), exclusive breastfeeding for at least 4 months and delaying the introduction of solid foods for at least 4 months prevents or delays atopic dermatitis, cow's milk allergy, and wheezing early in life. There is no evidence for a protective effect beyond 4 months.

Answer 6

Three immunodeficiency syndromes, immune dysregulation, polyendocrinopathy, enteropathy, X-linked (IPEX), autosomal recessive hyper IgE syndrome (AR-HIES), and Wiskott–Aldrich syndrome (WAS) are associated with increased risk of food allergies. IPEX is a rare, X-linked immunodeficiency. Patients often present in infancy with a triad of enteropathy, autoimmune endocrinopathy, and dermatitis. They develop recurrent infections as well as severe IgE-mediated food allergies. IPEX is caused by mutations in *FOXP3*, a gene crucial for the development of T_{reg} cells. Patients with IPEX lack T_{reg} cells.

AR-HIES is another rare immunodeficiency that presents in childhood with recurrent infections, chronic eczema, extremely high serum IgE levels, and eosinophilia. Mutations in *DOCK8* account for the majority of AR-HIES cases. The specific immunologic defect in DOCK8 deficiency is still under investigation. However, it is interesting that patients with autosomal dominant hyper IgE syndrome (AD-HIES), which is due to a mutation in *STAT3*, do not have an increased incidence of food allergies despite similarly high levels of both total and antigen-specific IgE. This suggests, once again, that IgE alone is not sufficient to cause food allergy.

WAS is an X-linked disease characterized by eczema, thrombocytopenia, T-cell dysfunction, and immune dysregulation. The fact that IPEX, DOCK8 deficiency, and WAS share the presence of skin inflammation is consistent with the idea that sensitization to food allergens can occur through the skin, as is likely to be the case in atopic dermatitis.

Case 13

Answer 1

The vast majority of food allergies are mediated by proteins that are recognized in their intact conformation by IgE antibodies or as small peptides associated with MHC molecules when seen by T-cell receptors. Class II MHC molecules present peptides with an optimal length of 18–20 amino acids. The elemental formulas contain only individual amino acids, which are too small to be antigenic and can therefore remain undetected by IgE antibodies or T-cell receptors. Note that hydrolyzed milk formulas contain peptides of milk protein and therefore retain the potential (albeit reduced) to be allergenic.

Answer 2

IL-5 is a critical cytokine for the terminal differentiation of eosinophils and their release from the bone marrow. Anti-IL-5 therapy can reduce the numbers of eosinophils in the peripheral blood and esophageal tissues. However, persistent inflammation can occur even in the absence of eosinophils. Tissue numbers of mast cells, basophils, NKT cells, and lymphocytes are all increased in this disease. IL-5 monoclonal therapy would not directly address the inflammation caused by these non-eosinophil cells.

Answer 3

Although eosinophilic esophagitis is primarily driven by food allergens, environmental aeroallergens may contribute to the inflammation in the esophagus. It is not uncommon for patients to experience worsening of symptoms in the middle of a pollen season. There are even some patients whose eosinophilic esophagitis is present only during their pollen allergy seasons. The mechanisms involved in this exacerbation may include both IgE-mediated responses and delayed-type hypersensitivity to swallowed aeroallergens.

Answer 4

A central conundrum in our current understanding of food allergy is that food-specific T_H2 immune responses to foods can drive such distinct and usually mutually exclusive patterns of food reactions. Some patients suffer primarily from the consequences of cellular infiltration and consequent tissue damage, as occurs in eosinophilic esophagitis, whereas others exhibit primarily symptoms of immediate hypersensitivity, including anaphylaxis, but show little or no tissue damage. In Buck's case, sensitivity to two of the offending foods—soy and oats—was demonstrated only by patch testing, consistent with a non-IgE-mediated mechanism and little risk of anaphylaxis. However, Buck also had evidence of IgE-mediated sensitivity to milk, eggs, and chicken, yet he had no history of anaphylactic reactions to these foods. The reasons that patients with eosinophilic esophagitis do not typically exhibit immediate hypersensitivity to food antigens despite the presence of food-specific IgE antibodies are not completely clear. The lack of acute reactions may be related to an absence of critical effector mechanisms, including an expanded and activated pool of mucosal mast cells in the lower gastrointestinal tract. In the future, separately defining and developing tests for the critical effector arms of food anaphylaxis (in addition to measuring IgE antibodies) may provide a better basis for distinguishing patients at risk of acute reactions from those likely to experience chronic inflammation.

Case 14

Answer 1

The mucociliary apparatus, in which a layer of mucus containing trapped inhaled particles is constantly swept upwards to the pharynx, is the first line of defense. Alveolar macrophages, innate effector cells residing in the air spaces of the lung, are also important early defenders against *A. fumigatus*. In chronic and invasive disease, adaptive immune mechanisms driven by T cells become operative.

Answer 2

This would be quite unlikely. The pathophysiologic pathways leading to the development of ABPA begin with IgE sensitization to *A. fumigatus*. The tendency to produce specific IgE antibodies to multiple environmental allergens is generally restricted to atopic individuals. In Josephine's case, the presence of allergen-specific IgE antibodies for molds, tree pollens, cat and dog dander, her asthma, and her response to allergy medications all indicate she is atopic. An exception to this rule occurs in a very small subset (2%) of patients with cystic fibrosis, who develop ABPA despite the absence of evidence of atopy.

Answer 3

In both allergic asthma and ABPA, T_H2-driven T-cell and B-cell responses result in *A. fumigatus*-specific IgE-mediated mast-cell activation and mediator release, eosinophilia, and eosinophil-mediated inflammation, which lead to chronic tissue inflammation and remodeling. In ABPA, there is failure to clear *A. fumigatus* from the airways. Persistent *A. fumigatus* antigen initiates a mixed immune response including the activation of innate effector cells, driving T_H2 and probably T_H17 responses leading to a mixed neutrophilic and eosinophilic infiltrate. These processes result in tissue damage (fibrosis and bronchiectasis), which are not induced in T_H2-driven allergic responses and therefore do not occur in asthma.

Answer 4

Clinical deterioration of asthma with pulmonary infiltrates, a positive skin test to *A. fumigatus*, elevated total IgE, *A. fumigatus*-specific IgE, and precipitating serum antibodies against *A. fumigatus* provided the criteria for diagnosis.

Answer 5

Stage 4 (steroid-dependent stage).

Case 15

Answer 1

Although Frank's early presentation had features that could be consistent with either the acute or subacute forms of hypersensitivity pneumonitis, he seems to have evolved toward a clinical syndrome most consistent with acute disease. Early on he experienced subtle symptoms of cough and mild dyspnea on exertion. Eventually he also experienced fever, had crackles on chest auscultation, and was found to have patchy infiltrates on chest X-ray, all consistent with the acute form.

Answer 2

The most common form of breathlessness triggered by environmental exposures is asthma. Therefore the diagnosis of asthma must be excluded before hypersensitivity pneumonitis can be considered in a patient presenting with the complaint of dyspnea (breathlessness). In Frank's case the failure to respond to bronchodilators provided a strong hint that he was not suffering from the reversible small-airway obstruction characteristic of asthma. In addition, the lack of physical findings (wheezing and prolonged expiratory phase) and the absence of obstructive physiology on spirometry together effectively exclude the diagnosis in Frank's case.

Answer 3

Treatment of hypersensitivity pneumonitis is a tremendous challenge. Because disease progression is driven by chronic recurrent exposure to an inciting antigen, consideration of the diagnosis along with identification of the antigen and development of effective avoidance strategies are critical. Glucocorticoids may have some benefit in dampening the symptoms of acute and subacute forms of the illness, but they are ineffective in controlling or reversing the fibrosis present in the chronic form. Once established, the fibrotic changes cause a fixed loss of lung function and are, unfortunately, irreversible. Cigarette smoke may accelerate disease progression, and smokers should be counseled regarding cessation.

Answer 4

CD8 T cells predominate in the BAL of patients with hypersensitivity pneumonitis. This is suggestive of a type IV, delayed-type hypersensitivity mechanism in which antigen-specific T cells elaborate cytokines and chemokines that orchestrate a chronic inflammatory response in the lungs. γ:δ T cells may also be present in the BAL, suggesting that lipid antigens may have a role. The occasional recovery of connective-tissue-type mast cells and macrophages from BAL is consistent with the fibrosis that occurs in chronic disease. Both cell types are known to drive fibrosis.

Case 16

Answer 1

Hymenopteran venoms can result in direct and spontaneous activation of mast cells and basophils, resulting in systemic reactions in patients lacking detectable venom-specific IgE antibodies.

Answer 2

Peripheral blood mononuclear cells (PBMCs), comprising both antigen-presenting cells and T cells, from a patient allergic to bees will proliferate when cultured with a major honeybee allergen, such as phospholipase A2 (PLA2). Therefore, one could isolate PMBCs from a bee-allergic patient who is also undergoing VIT, before initiation of VIT and after 7, 14, and 28 days of immunotherapy. These PBMCs can be stimulated with PLA2 in the presence and absence of neutralizing anti-IL-10 antibodies for 7 days. Lymphocyte proliferation can be measured at the end of the 7-day incubation period using ^{3}H-thymidine incorporation, a measure of DNA synthesis and hence cell replication. If the neutralizing anti-IL-10 antibodies result in increased lymphocyte proliferation compared with cultures lacking the anti-IL-10 antibodies, one can infer that IL-10 decreases lymphocyte proliferation in response to PLA2 and thus mediates immunologic tolerance.

Answer 3

It is thought that the allergen-specific blocking antibodies, most commonly of the IgG4 subtype, bind to the venom protein before the allergen can bind to the IgE found on mast cells (Fig. A16.3). IgE antibodies enhance the T-cell responses by facilitating endocytosis of proteins by antigen-presenting cells via mechanisms involving both the low-affinity IgE receptor CD23 and the high-affinity receptor FcεRI. Blocking IgG4 antibodies can also inhibit this IgE-mediated allergen presentation to T cells.

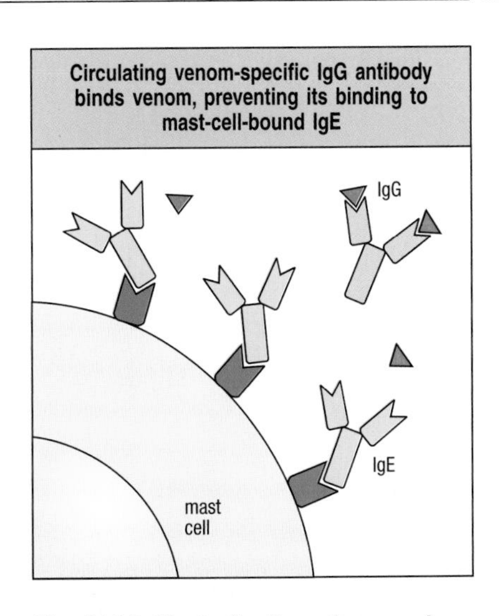
Circulating venom-specific IgG antibody binds venom, preventing its binding to mast-cell-bound IgE
IgG
IgE
mast cell

Fig. A16.3 Neutralization of venom by circulating IgG antibodies prevents activation of mast cells by venom binding to cell-bound IgE.

Answer 4

No. Neither the size of a skin-test reaction nor the level of venom-specific antibodies is a good predictor of the severity of a patient's hypersensitivity reaction. Patients with minimal levels of IgE antibodies against venoms or mild skin-test positivity are still at risk of severe hypersensitivity reactions to venoms. However, as discussed for Alec, these patients should also have a baseline tryptase level taken to exclude a diagnosis of mastocytosis.

Answer 5

Some patients with positive testing to multiple venoms will have a true hypersensitivity to antigens specific to each venom. Other patients will have IgE specific to one venom that cross-reacts with similar carbohydrate epitopes of glycoproteins found in another venom.

Answer 6

A patient with mastocytosis has a higher burden of mast cells throughout his or her body. Because insect venoms can activate mast cells directly in an IgE-independent manner, these patients are at risk of a massive release of histamine or the other vasoactive mediators that cause anaphylactic reactions.

Case 17

Answer 1

Eosinophils initially infiltrate the myocardium but can subsequently cause inflammation and necrosis. This can lead to the generation of a mural thrombus adjacent to the site of necrosis as well as to fibrosis of the myocardium. Once fibrosis occurs it is irreversible. For that reason, cardiac involvement needs to be treated immediately. For patients with ongoing eosinophilia, a periodic check for cardiac involvement is recommended—for example, by measuring troponin levels or performing echocardiography.

Answer 2

Perhaps the activation state of the eosinophils determines the amount of tissue damage. The more activated the eosinophil, the greater the release of toxic mediators and the more inflammation and tissue damage that will occur. Quiescent eosinophils might infiltrate tissue without causing damage.

Answer 3
Pediatric hypereosinophilia is often associated with acute lymphoblastic leukemia, a common childhood cancer. HES without associated malignancy is uncommon in children, but some cases do occur. For unknown reasons, the *PDGFRα-FIP1L1* fusion gene is rare in pediatric HES.

Answer 4
Strongyloides infestation is responsible for some cases of eosinophilia. Treatment with corticosteroids can lead to widespread dissemination of this parasite, often called hyperinfection, which can be fatal. Screening should prevent this.

Answer 5
Many cancers and myeloproliferative diseases are characterized by somatic mutations leading to tyrosine kinase hyperactivation. Imatinib was initially developed to block the activity of the Bcr–Abl fusion protein, encoded on the so-called Philadelphia chromosome, in chronic myelogenous leukemia. In addition, mastocytosis, an abnormal proliferation of mast cells, is caused by point mutations in the *KIT* gene that lead to constitutive activity.

Answer 6
Chronic eosinophilia of the lung is associated with allergic asthma. Inhaled corticosteroids are the most effective treatment for persistent asthma, and studies have shown that titrating their dose on the basis of sputum eosinophil counts improves asthma control. These data suggest that anti-IL-5 antibody might be effective in asthma. Recent studies have indicated that mepolizumab is effective for a subset of patients with severe asthma and high lung levels of eosinophils and can significantly increase the control of asthma in these patients. It seems less effective in milder asthma.

Case 18

Answer 1
Churg–Strauss syndrome is usually associated with adult-onset asthma. Allergic asthma most commonly begins in early childhood, with about 75% of patients diagnosed before the age of 7 years. But in longitudinal studies, many patients diagnosed in adulthood actually had symptoms of wheezing or bronchial hyperresponsiveness much earlier, suggesting a true childhood onset. Most patients with allergic asthma have a significant family history of atopy; this is not true for Churg–Strauss disease.

Answer 2
With such predominant eosinophilia, IL-5 would be a reasonable target for therapy. Mepolizumab, a monoclonal antibody against IL-5, has been successful in treating hypereosinophilic syndrome. Preliminary studies of mepolizumab in CSS have been promising. Because TNF-α production is important for the development of granulomas, TNF-α antagonists have also been tried for Churg–Strauss vasculitis. Unfortunately, although TNF-α antagonists have proved useful for other types of vasculitis, etanercept (a TNF-α receptor fusion protein) has not shown significant benefit in patients with CSS.

Answer 3
The results of Georgiana's pulmonary function tests would probably be similar to those in asthmatic patients. She would show an obstructive defect, with disproportionately reduced forced expiratory volume. Because pulmonary

vasculitis can also cause interstitial lung disease and reduction of gas exchange, her diffusion lung capacity might also be decreased.

Answer 4
Hypereosinophilic syndrome (HES) also presents with overwhelming eosinophil infiltration to various organs and tissues and can have many of the same clinical features. However, vasculitis is not a typical component of HES. The skin findings in HES are more similar to those seen with allergic disease, and biopsies would not show vasculitis.

Answer 5
Rituximab targets and depletes B cells. B-cell depletion ameliorates CSS in part by lowering autoantibody levels, including that of the pathogenic ANCA. However, rituximab targets CD20, which is not expressed on plasma cells (the primary source of antibody production). The effect of rituximab on ANCA titers can therefore be variable. B cells may also contribute to autoimmune disease by providing co-stimulatory help to T cells and secreting inflammatory cytokines. Amelioration of disease by B-cell depletion may thus involve multiple mechanisms.

Case 19

Answer 1
Aspirin-hypersensitive patients react to all NSAIDs that preferentially inhibit COX1. In contrast, selective COX2 inhibitors, given in therapeutic doses, are well tolerated by aspirin-sensitive individuals. Acetaminophen (Tylenol) is a weak inhibitor of COX1, and most patients with AERD can tolerate standard doses of it. Administration of higher than normal doses of acetaminophen has been associated with respiratory reactions.

Answer 2
Some clinical features seen in anaphylaxis can also occur in aspirin-hypersensitivity reactions, such as bronchospasm and excessive mucus production in the upper respiratory tract. Alterations in the synthesis of eicosanoids are directly responsible for most of the effects observed in aspirin sensitivity, but they also have a major role in IgE-mediated reactions because mast cells are triggered to synthesize Cys-LTs after cross-linking of the IgE receptor, FcεRI. Together, the three Cys-LTs—LTC_4, LTD_4, and LTE_4—constitute the so-called 'slow-reacting substance of anaphylaxis' (SRS-A), a substance that was found to be responsible for smooth muscle contraction and vascular dilation in anaphylaxis decades before the individual leukotrienes were identified. (Mast-cell activation also occurs during aspirin-provoked reactions, as evidenced by increased serum levels of mast-cell mediators such as tryptase, histamine, and metabolites of PGD_2.)

Answer 3
Elevated levels of leukotriene metabolites have been measured in body fluids of patients with AERD. Levels of urinary LTE_4 are distinctly elevated in aspirin-hypersensitive patients at baseline and rise further in a time-dependent manner after aspirin exposure, coinciding with aspirin-induced respiratory reactions. However, the diagnostic reliability of increased urinary excretion of LTE_4 has been questioned, and it is not currently used in clinical practice.

Answer 4
Montelukast (a $CysLT_1$ receptor antagonist) and zileuton (a 5-lipoxygenase antagonist) are used in the treatment of asthma. They are also commonly used

in patients with AERD, but their clinical utility is limited because they only partly inhibit aspirin-provoked symptoms. However, leukotriene antagonists are particularly useful in preventing lower respiratory symptoms during aspirin challenges or desensitizations, and they are routinely used in this setting. It should be noted that asthma in aspirin-sensitive patients does not respond better to treatment with leukotriene antagonists than does asthma in aspirin-tolerant patients.

Case 20

Answer 1

Mastocytosis is one of the conditions that needs to be considered in the differential diagnosis of patients with idiopathic anaphylaxis. Symptoms such as urticaria, angioedema, and wheezing are common in patients with idiopathic or allergen-induced anaphylaxis but are not commonly observed in mast-cell degranulation episodes due to mastocytosis. The basis of this difference in symptomatology is not well understood. Symptoms of flushing, lightheadedness, abdominal cramps, nausea, vomiting, and diarrhea are common in both conditions.

Answer 2

Drugs such as interferon-α and cladribine may result in a partial and temporary reduction in mast cell numbers but can have serious adverse effects, including the suppression of bone marrow function. Considering that most patients with indolent mastocytosis have a normal life expectancy, the risk-versus-benefit analysis currently favors conservative symptomatic treatment rather than cytoreduction for these patients.

Answer 3

Imatinib inhibits normal Kit by binding to the active site of the tyrosine kinase domain. The structure of the active site is altered by the D816V mutation commonly present in patients with mastocytosis such that it is no longer accessible to imatinib. Therefore imatinib is not a good inhibitor for most patients with mastocytosis, who carry the D816V mutation. Occasional patients with aggressive systemic mastocytosis caused by mutations outside codon 816 or mutations that generate other targets of imatinib (such as rearrangements involving the receptors for platelet-derived growth factor or the Bcr–Abl fusion) can be good candidates for this drug.

Answer 4

Systemic mastocytosis is associated with osteoporosis in a subset of patients. Therefore all adult patients should be screened by bone densitometry for the onset of osteoporosis. Gastric and duodenal ulcers and gastritis may develop as a result of hypersecretion of gastric acids induced by histamine released from mast cells. Because systemic mastocytosis usually represents a disorder of a hematopoietic progenitor cells in the bone marrow, patients should be monitored for the development of other hematologic disorders such as myeloid leukemias, myelodysplastic syndromes, and lymphoproliferative disorders. Although about 10–20% of patients have these other non-mast-cell hematologic disorders at the time of their diagnosis of mastocytosis, the risk of developing such disorders as a progression from cutaneous or indolent mastocytosis is fortunately smaller (approximately 5% or less), although higher than in the general population.

Answer 5

In people sensitized to hymenopteran venom, a sting can activate mast cells via IgE. Hymenopteran venoms also contain compounds, such as mastoparan, that can activate mast cells directly, independently of IgE. It is thought that the IgE- and non-IgE-mediated activation properties of the venoms, combined with the increased mast-cell burden and an intrinsic reduction in the mast-cell activation threshold imparted by the D816V mutation, interact to increase the risk of systemic anaphylactic reactions in patients with mastocytosis or monoclonal mast-cell activation syndrome. Tests for mastocytosis should therefore be considered for patients with systemic reactions to hymenopteran venoms. Those who are found to have mast-cell disease should be considered for lifelong immunotherapy with the venoms to which they are sensitized. It should be noted that patients with mastocytosis may also develop systemic reactions to venom immunotherapy. For some patients it may be necessary to modify the immunotherapy protocols by premedication with drugs that target mast-cell mediators, and to start with lower venom doses.

Answer 6

Giles's symptoms of episodic facial flushing associated with tachycardia, light-headedness, abdominal symptoms, and near syncope are, like an episode of anaphylaxis, caused by the release of vasoactive mast-cell mediators and are similarly responsive to epinephrine. They can be triggered by heat, stress, drugs, and hymenopteran stings, and can also occur without a clear trigger. Patients with mastocytosis may also experience anaphylactic reactions even when they have no history of anaphylaxis. Therefore the risk-to-benefit ratio favors prescribing epinephrine to all patients diagnosed with mastocytosis.

Figure Acknowledgments

Case 3
Fig. 3.2 adapted from *Janeway's Immunobiology,* 8th edition, by Kenneth Murphy. © 2012 by Garland Science. Used by permission of Garland Science.
Fig. 3.3 from Cameron, L., Gounni, A.S., Frenkiel, S. et al.: SεSμ and SεSγ switch circles in human nasal mucosa following *ex vivo* allergen challenge: evidence for direct as well as sequential class switch recombination. *Journal of Immunology.* 2003, **171**: 3816–3822. © 2003 The American Association of Immunologists. With permission from The American Association of Immunologists.
Fig. 3.4 adapted from Kagan, S., Lewis, W. and Levetin, E.: Aeroallergen PhotoLibrary. http://allernet.net/photolibrary.html © 2004.
Fig. 3.6 panel b adapted from *Janeway's Immunobiology,* 8th edition, by Kenneth Murphy. © 2012 by Garland Science. Used by permission of Garland Science.

Case 4
Fig. 4.1 from *Janeway's Immunobiology,* 8th edition by Kenneth Murphy. © 2012 by Garland Science. Used by permission of Garland Science.
Fig. 4.2 from *Janeway's Immunobiology,* 8th edition, by Kenneth Murphy. © 2012 by Garland Science. Used by permission of Garland Science.
Fig. 4.4 from *Middleton's Allergy: Principles and Practice*, 7th edition. Eds. Adkinson, N.F., Bochner, B.S., Busse, W.W. et al. © 2009 Elsevier. With permission from Elsevier.
Fig. 4.5 from Yichieh, S., Colby, K. and Dohlman, C.: 19 year old man with a "corneal abrasion." *Digital Journal of Ophthalmology.* 1997; Harvard Medical School, vol 3, no. 15. With permission from Massachusetts Eye And Ear Infirmary.

Case 5
Fig. 5.4 from Leung, D.Y.M., Bhan, A.K., Schneeberger, E.E. and Geha, R.S.: Characterization of the mononuclear cell infiltrate in atopic dermatitis using monoclonal antibodies. *Journal of Allergy and Clinical Immunology.* 1983; **71**: 47–56. © 1983 American Academy of Allergy. With permission from Mosby.

Case 6
Fig. 6.3 from http://hardinmd.lib.uiowa.edu/. © 2007 by Interactive Medical Media LLC. All rights reserved.

Case 7
Fig. 7.2 adapted from *Janeway's Immunobiology,* 8th edition, by Kenneth Murphy. © 2012 by Garland Science. Used by permission of Garland Science.

Case 9
Fig. 9.1 from Kamradt, T. and Mitchison, N.A.: Advances in Immunology: Tolerance and Autoimmunity. *New England Journal of Medicine.* 2001; **344**: 655–664. © 2001 Massachusetts Medical Society. With permission from the Massachusetts Medical Society. All rights reserved.

Case 10
Fig. 10.5 from Williams, K.: Hereditary angioneurotic edema. *Clinical Immunology.* 1975; **4**: 174–188. © 1975 Elsevier. With permission from Elsevier.

Case 11
Fig. 11.3 from *Robbins Basic Pathology*, 8th edition, by Vinay Kumar. © 2007 Elsevier. With permission from Elsevier.

Case 12
Fig. 12.3 adapted from Sampson, H.A. and Ho, D.G.: Relationship between food-specific IgE concentrations and the risk of positive food challenges in children and adolescents. *Journal of Allergy and Clinical Immunology.* 1997; **100**: 444–451. © 1997 American Academy of Allergy. With permission from Mosby.
Fig. 12.4 adapted from Sicherer, S.H. and Sampson, H.A.: Food allergy. *Journal of Allergy and Clinical Immunology.* 2010; **125**: S116–S125. © 2010 American Academy of Allergy. With permission from Mosby.

Case 13
Fig. 13.1 adapted from *Janeway's Immunobiology,* 8th edition, by Kenneth Murphy. © 2012 by Garland Science. Used by permission of Garland Science.

Case 14
Fig. 14.4 from Rosenberg, M., Patterson, R., Mintzer, R. et al.: Clinical and Immunologic Criteria for the Diagnosis of Allergic Bronchopulmonary Aspergillosis. *Annals of Internal Medicine.* 1977; **86**: 405–414. © 1977 American College of Physicians. With permission from American College of Physicians.
Fig. 14.5 from Agarwal, R.: Allergic bronchopulmonary aspergillosis. *Chest.* 2009; **135**: 805–826. © 2009 American College of Chest Physicians. With permission from the American College of Chest Physicians.
Fig. 14.6 from Agarwal, R.: Allergic bronchopulmonary aspergillosis. *Chest.* 2009; **135**: 805–826. © 2009 American College of Chest Physicians. With permission from the American College of Chest Physicians.

Case 15
Fig. 15.2 from Hanak, V., Kalra, S., Aksamit, T.R. et al.: Hot tub lung: presenting features and clinical course of 21 patients. *Respiratory Medicine.* 2006; **100**: 610–615. © 2006 Elsevier. With permission from Elsevier.
Fig. 15.3 from Patel, A.M., Ryeu, J.H. and Reed, C.H.: Hypersensitivity pneumonitis: current concepts and future questions. *Journal of Allergy and Clinical Immunology.* 2001; **108**: 661–670. © 2001 American Academy of Allergy. With permission from Mosby.
Fig. 15.5 from Hanak, V., Golbin, J.M. and Ryu, J.H.: Causes and presenting features in 85 consecutive patients with hypersensitivity pneumonitis. *Mayo Clinic Proceedings.* 2007; **82**: 812–816 © 2007 Elsevier. With permission from Elsevier.
Fig. 15.6 from Hanak, V., Kalra, S., Aksamit, T.R. et al.: Hot tub lung: presenting features and clinical course of 21 patients. *Respiratory Medicine.* 2006; **100**: 610–615. © 2006 Elsevier. With permission from Elsevier.

Case 16

Fig. 16.1 from *Janeway's Immunobiology,* 8th edition by Kenneth Murphy. © 2012 by Garland Science. Used by permission of Garland Science.

Fig. 16.4 adapted from Insect Allergy, by Golden, D.B.K., in *Middleton's Allergy: Principles and Practice*, 7th edition, vol 2. 2009. Edited by Adkinson, N.F., Bochner, B.S., Busse, W.W. et al. © 2009 Elsevier. With permission from Elsevier.

Fig. 16.5 from Texas Imported Fire Ant Research and Management Project. http://fireant.tamu.edu/materials/multimedia_photos/index.php. With permission from the Texas Department of Agriculture.

Fig. A16.3 from *Janeway's Immunobiology,* 8th edition by Kenneth Murphy. © 2012 by Garland Science. Used by permission of Garland Science.

Case 17

Fig. 17.2 from *Janeway's Immunobiology,* 8th edition by Kenneth Murphy. © 2012 by Garland Science. Used by permission of Garland Science.

Fig. 17.4 from Page, K. and Zenilman, J.: Eosinophilia in a patient from South America. *The Journal of the American Medical Association.* 2008; **299**: 437–444. © 2008 American Medical Association. With permission from the American Medical Association.

Fig. 17.5 from Cools, J., DeAngelo, D.J., Gotlib, J. et al.: A tyrosine kinase created by fusion of the *PDGFRA* and *FIP1L1* genes as a therapeutic target of imatinib in idiopathic hypereosinophilic syndrome. *New England Journal of Medicine.* 2003; **348**: 1201–1214. © 2003 Massachusetts Medical Society. With permission from the Massachusetts Medical Society. All rights reserved.

Fig. 17.6 from Fazel, R., Dhaliwal, G., Saint, S. and Nallamothu, B.K.: Clinical problem-solving. A red flag. *New England Journal of Medicine.* 2009; **360**: 2005–2010. © 2009 Massachusetts Medical Society. With permission from the Massachusetts Medical Society. All rights reserved.

Case 18

Fig. 18.2 from Masi, A.T., Hunder, G.G., Lie, J.T. et al.: The American College of Rheumatology 1990 criteria for the classification of Churg-Strauss syndrome (allergic granulomatosis and angiitis). *Arthritis Rheumatology.* 1990; **33**: 1094–1100. © 1990 John Wiley & Sons. With permission from John Wiley & Sons.

Case 19

Fig. 19.4 panel a from Hawke, M. (1993) *Pocket Atlas of the Nose and Paranasal Sinuses.* http://hawkelibrary.com/ With permission from Hawke, M.

Case 20

Fig. 20.1 adapted from *Janeway's Immunobiology,* 8th edition by Kenneth Murphy. © 2012 by Garland Science. Used by permission of Garland Science.

Fig. 20.3 from Brockow, K., Akin, C., Huber, M. and Metcalfe, D.D. Assessment of the extent of cutaneous involvement in children and adults with mastocytosis: relationship to symptomatology, tryptase levels, and bone marrow pathology. *Journal of the American Academy of Dermatology.* 2003; **48**: 508–516. © 2003 American Academy of Dermatology. With permission from Mosby.

Fig. 20.5 data from Horny, H-P., Akin, C., Metcalfe, D.D. et al. in *WHO Classification of Tumours of Haematopoietic and Lymphoid Tissues* vol 2, 2008, p 54. Edited by Swerdlow, S.H., Campo, E., Harris, N.L. IARC Press. © 2008 World Health Organization.

Fig. 20.7 data from Horny, H-P., Akin, C., Metcalfe, D.D. et al. in *WHO Classification of Tumours of Haematopoietic and Lymphoid Tissues* vol 2, 2008, p 54. Edited by Swerdlow, S.H., Campo, E., Harris, N.L. IARC Press. © 2008 World Health Organization.

Index

Note: Figures for Case numbers 1 to 20 are labeled in the form **Fig. 1.1**, **Fig 2.1** etc., and those in the Answer section are labeled by Case number, in the form **Fig. A16.3.**

A

D

E

F

K

L

M